Algorithmic Aspects of Flows in Networks

by

Günther Ruhe

Department of Mathematics and Computer Science,
Leipzig University of Technology,
Leipzig, Germany

KLUWER ACADEMIC PUBLISHERS
DORDRECHT / BOSTON / LONDON

Library of Congress Cataloging-in-Publication Data

Ruhe, Günther.
Algorithmic aspects of flows in networks / by Günther Ruhe.
p. cm. -- (Mathematics and its applications ; v. 69)
Includes bibliographical references (p.) and index.
ISBN 0-7923-1151-5 (hb : acid-free paper)
1. Network analysis (Planning) I. Title. II. Series:
Mathematics and its applications (Kluwer Academic Publishers) ; v. 69.
T57.85.R84 1991
658.4'032--dc20 91-2594

ISBN 0-7923-1151-5

Published by Kluwer Academic Publishers,
P.O. Box 17, 3300 AA Dordrecht, The Netherlands.

Kluwer Academic Publishers incorporates
the publishing programmes of
D. Reidel, Martinus Nijhoff, Dr W. Junk and MTP Press.

Sold and distributed in the U.S.A. and Canada
by Kluwer Academic Publishers,
101 Philip Drive, Norwell, MA 02061, U.S.A.

In all other countries, sold and distributed
by Kluwer Academic Publishers Group,
P.O. Box 322, 3300 AH Dordrecht, The Netherlands.

Printed on acid-free paper

Printed in the Netherlands

SERIES EDITOR'S PREFACE

'Et moi, ..., si j'avait su comment en revenir, je n'y serais point allé.'

Jules Verne

The series is divergent; therefore we may be able to do something with it.

O. Heaviside

One service mathematics has rendered the human race. It has put common sense back where it belongs, on the topmost shelf next to the dusty canister labelled 'discarded non-sense'.

Eric T. Bell

Mathematics is a tool for thought. A highly necessary tool in a world where both feedback and non-linearities abound. Similarly, all kinds of parts of mathematics serve as tools for other parts and for other sciences.

Applying a simple rewriting rule to the quote on the right above one finds such statements as: 'One service topology has rendered mathematical physics ...'; 'One service logic has rendered computer science ...'; 'One service category theory has rendered mathematics ...'. All arguably true. And all statements obtainable this way form part of the raison d'être of this series.

This series, *Mathematics and Its Applications*, started in 1977. Now that over one hundred volumes have appeared it seems opportune to reexamine its scope. At the time I wrote

> "Growing specialization and diversification have brought a host of monographs and textbooks on increasingly specialized topics. However, the 'tree' of knowledge of mathematics and related fields does not grow only by putting forth new branches. It also happens, quite often in fact, that branches which were thought to be completely disparate are suddenly seen to be related. Further, the kind and level of sophistication of mathematics applied in various sciences has changed drastically in recent years: measure theory is used (non-trivially) in regional and theoretical economics; algebraic geometry interacts with physics; the Minkowsky lemma, coding theory and the structure of water meet one another in packing and covering theory; quantum fields, crystal defects and mathematical programming profit from homotopy theory; Lie algebras are relevant to filtering; and prediction and electrical engineering can use Stein spaces. And in addition to this there are such new emerging subdisciplines as 'experimental mathematics', 'CFD', 'completely integrable systems', 'chaos, synergetics and large-scale order', which are almost impossible to fit into the existing classification schemes. They draw upon widely different sections of mathematics."

By and large, all this still applies today. It is still true that at first sight mathematics seems rather fragmented and that to find, see, and exploit the deeper underlying interrelations more effort is needed and so are books that can help mathematicians and scientists do so. Accordingly MIA will continue to try to make such books available.

If anything, the description I gave in 1977 is now an understatement. To the examples of interaction areas one should add string theory where Riemann surfaces, algebraic geometry, modular functions, knots, quantum field theory, Kac-Moody algebras, monstrous moonshine (and more) all come together. And to the examples of things which can be usefully applied let me add the topic 'finite geometry'; a combination of words which sounds like it might not even exist, let alone be applicable. And yet it is being applied: to statistics via designs, to radar/sonar detection arrays (via finite projective planes), and to bus connections of VLSI chips (via difference sets). There seems to be no part of (so-called pure) mathematics that is not in immediate danger of being applied. And, accordingly, the applied mathematician needs to be aware of much more. Besides analysis and numerics, the traditional workhorses, he may need all kinds of combinatorics, algebra, probability, and so on.

In addition, the applied scientist needs to cope increasingly with the nonlinear world and the extra mathematical sophistication that this requires. For that is where the rewards are. Linear models are honest and a bit sad and depressing: proportional efforts and results. It is in the non-linear world that infinitesimal inputs may result in macroscopic outputs (or vice versa). To appreciate what I am hinting at: if electronics were linear we would have no fun with transistors and computers; we would have no TV; in fact you would not be reading these lines.

There is also no safety in ignoring such outlandish things as nonstandard analysis, superspace and anticommuting integration, p-adic and ultrametric space. All three have applications in both electrical engineering and physics. Once, complex numbers were equally outlandish, but they frequently proved the shortest path between 'real' results. Similarly, the first two topics named have already provided a number of 'wormhole' paths. There is no telling where all this is leading - fortunately.

Thus the original scope of the series, which for various (sound) reasons now comprises five subseries: white (Japan), yellow (China), red (USSR), blue (Eastern Europe), and green (everything else), still applies. It has been enlarged a bit to include books treating of the tools from one subdiscipline which are used in others. Thus the series still aims at books dealing with:

- a central concept which plays an important role in several different mathematical and/or scientific specialization areas;
- new applications of the results and ideas from one area of scientific endeavour into another;
- influences which the results, problems and concepts of one field of enquiry have, and have had, on the development of another.

Networks occur everywhere: networks of roads, networks for information flow in management and elsewhere, computer networks, transport networks, ... In addition many other problems such as planning a complex project, machine scheduling, production management, manpower planning, ..., can be cast in network form.

In all these cases the maximal flow through the network is of prime importance and determining that flow is crucial. And hence so are the algorithms for doing so. It is therefore no surprise that all sorts of questions concerning such algorithms attract a lot of research interest: theoretical foundations, questions of efficiency and implementation, questions of complexity and design, etc. These traditional matters together with modern developments such as parametric and multicriteria flows form the subject matter of this monograph.

Many algorithms are given and analysed in detail. Both for the relative novice and for the expert this seems to me to be a thorough and up-to-date text on the algorithmic aspects of networks. In virtually all fields of mathematics there is a smaller or larger gap between the established body of knowledge and actually calculating solutions in concrete cases. For networks this book achieves a great deal in bridging this gap.

The shortest path between two truths in the real domain passes through the complex domain.

J. Hadamard

La physique ne nous donne pas seulement l'occasion de résoudre des problèmes ... elle nous fait pressentir la solution.

H. Poincaré

Never lend books, for no one ever returns them; the only books I have in my library are books that other folk have lent me.

Anatole France

The function of an expert is not to be more right than other people, but to be wrong for more sophisticated reasons.

David Butler

Bussum, 18 February 1991

Michiel Hazewinkel

CONTENTS

INTRODUCTION

Flows in networks are of growing interest from the point of view of both theory and applications. The development of very efficient algorithms for most classes of network problems combined with the powerful computers now available have led to an increasing number of applications. The large spectrum of real-world applications includes production-distribution, urban traffic, manpower planning, computer networks, facility location, routing, and scheduling. Starting from fundamental models as maximum flows or minimum-cost flows, more advanced models have been investigated. Related to theory, network flows have been proven to be an excellent indicator of things to come in other areas , especially in mathematical programming.

There is a common interest from theory and applications in the study of algorithms. Network flow algorithms with their theoretical foundations and questions of the design and the analysis of efficient computer implementations are the main subject of this book. An overview of recent algorithmic results and developments for the fundamental problem classes is given. Moreover, advanced topics such as multicriteria and parametric flows or the detection of network structure in large scale linear programs and solution procedures for problems with embedded network structure are presented. In all these cases, the emphasis is laid on algorithmic aspects. Considering the study of algorithms as the bridge between theory and applications, the book will contribute to both areas.

In the first chapter, necessary foundations are summarized. This concerns some general notations, preliminaries from graph theory and computational complexity. The latter subjects, in combination with linear and integer programming, constitute the main source of network optimization. Chapters 2 - 4 are devoted to maximum flows, minimum cost flows and flows in generalized networks. For each of these models, recent theoretical and algorithmic results are summarized. Additionally, one algorithm is described in more detail, each. For the description of the algorithms we use a Pascal-like pseudo-code that can easily be understood. For determining maximum flows, the preflow algorithm of Goldberg (1985) and the different implementation variants studied by Derigs & Meier (1989) are investigated. In Chapters 3 and 4, the fundamental simplex algorithm is applied to the minimum-cost flow problem in pure and generalized networks, respectively. Computational results obtained on randomly generated test examples give some information on the dependence between CPU-time and problem parameters. For the generalized circulation problem, the recently developed first combinatorial algorithm of Goldberg, Plotkin & Tardos (1988) is described.

Chapters 5 - 8 are devoted to more advanced models. The possibility to handle them is based on the efficiency achieved in solving the problem classes of the previous chapters.

The majority of problems appearing in practice cannot be analyzed adequately without taking into account more than one criterion. In §5, solution approaches and algorithms for multicriteria flow problems are investigated. Special emphasis is laid on bicriteria minimum-cost flow problems. Based on an exponential dependence between the number of vertices and the number of efficient extreme points in the objective space exact as well as approximate methods are considered. Especially, the notion of efficiency or Pareto-optimality is relaxed leading to ϵ-optimality. This new concept is implemented on a computer and allows to fix the number of efficient solutions and to give a bound on the relative improvement which is yet possible related to the calculated set. On the other side, for an a priori accuracy the algorithm computes the necessary number of solutions to achieve this level. Numerical results for NETGEN-generated test examples are given. As an application, the optimal computer realization of linear algorithms is modeled and solved as a bicriteria shortest path problem.

In the case of dynamic settings, successive instances of problems differing only in the modification of some parts of the problem data, must be solved. These modifications may occur in the capacity constraints and in the objective function. §6 is devoted to parametric flows in networks. Again, a careful analysis of the inherited complexity is the base of the subsequent algorithmic investigations. This is done with special emphasis on the number of breakpoints in the optimal value function. For the parametric maximum flow problem, the piece-wise or vertical algorithm is studied. Klinz (1989) implemented the algorithm. The computational results including a comparison with other algorithms to solve this problem are reported in Section 6.3. For generalized networks, the horizontal solution method of Hamacher & Foulds (1989), originally developed for pure networks, can be applied. Dual reoptimization procedures are a promising tool to solve parametric problems. The parametric dual network simplex method and its realization for the minimum-cost flow problem are studied in detail in Section 6.5. Finally, network flows with fuzzy data are handled. The near equivalence between fuzzy data and parametric programming is demonstrated for the case of fuzzy maximum flows.

Optimization algorithms that exploit network structure can be up to one-hundred times faster than general, matrix-oriented LP-algorithms. This is due to the use of linked lists, pointers and logical instead of arithmetic operations. The possibility of displaying the graph-theoretical structure in two-dimensional drawings greatly simplifies the insight into the problem and the interpretation of the results. With the sophisticated graphical equipment of today, this fact leads to a higher acceptance of network-related problem formulations. The aim in detecting network structure in linear programming is to make as much use of the above-mentioned advantages of network-related problem formulations as possible. In addition to studying complexity results, two different approaches to extract network structure are discussed. In the case of detecting hidden networks, computational results for an implemented procedure are given.

The final Chapter 8 investigates the solution of network flow problems with additional linear constraints. In other words, we have a linear programming problem with an embedded network structure where the network should be as large as possible. It is demonstrated that by exploiting the special structure of this kind of problems, the algorithms can be made more efficient than general linear programming. As an application, the transformation of an interval scheduling problem to a network flow problem with side constraints and its solution via surrogate constraints and parametric programming is demonstrated.

This book is a synthesis of a reference and a textbook. It is devoted to graduate students in the fields of operations research, mathematics, computer science and also in engineering disciplines. Since recent developments in both theory and algorithms are studied, the book can also be used for researchers and practitioners. Both can use the book to learn about the existing possibilities in solving network flow optimization problems.

The research topics covered in Chapters 5 and 6 were greatly stimulated by the scientific cooperation between the Department of Mathematics of Technical University Graz and the Department of Mathematics and Computer Science of Leipzig University of Technology on "Lösungsmethoden für multikriterielle Entscheidungsprobleme bei Netzplänen und Reihenfolgeproblemen". Part of the work was done while the author visited Universität Augsburg. This stay was supported by a Grant from Deutsche Forschungsgemeinschaft on "Anwendungsbezogene Optimierung und Steuerung".

The author greatly appreciates the suggestions and remarks given by Bettina Klinz and Roland Kopf who read parts of the book and contributed to some of the computer programs. Jörg Sawatzki implemented the algorithm of Section 2.7. Finally, I would like to thank my family for their patience during the preparation of the book.

Leipzig, January 1990 Günther Ruhe

§1 FOUNDATIONS

1.1. General Preliminaries

The symbols **N**, **Z**, and **R** are used for the sets of positive integers, integers and real numbers, respectively. The n-dimensional real vector space, the set of ordered n-tuples of real numbers, is denoted by $\mathbf{R}^n$. The inner product of two vectors $x,y \in \mathbf{R}^n$ is denoted by x^Ty. If x is a real number, then $\lfloor x \rfloor$ and $\lceil x \rceil$ denote the lower integer part respectively the upper integer part of x.

The set of elements $e_1,e_2,...$ is written explicitly as E = $\{e_1,e_2,...\}$. The second possibility to represent a set is by defining all elements for which a condition C is fulfilled: E = {e: C(s) is true}. Given a set A, we denote by #(A) the cardinality of A.

For two sets A,B we write

$A \subset B$ if A is contained in B (and possibly A = B);
$A \subsetneq B$ if A is strictly contained in B;
B - A for the elements in the set B which do not belong to A;
A + B for the union of A and B;
$A \cap B$ for the intersection of A and B.

An optimization problem is to minimize some objective function f(x) defined on a feasible area $\mathcal{S}$. This is written as

(1) $\min \{f(x): x \in \mathcal{S}\}$.

The notation $x^* \in \arg\min \{f(x): x \in \mathcal{S}\}$ means that x^* is a minimum solution of (1).

A mapping μ from a set S to a set T is written as $\mu: S \longmapsto T$.

For a given matrix A, det(A) denotes its determinant. Finally, we use the following standard functions:

(2) $$\text{abs}(x) := \begin{cases} x & \text{for } x \geq 0 \\ -x & \text{for } x < 0 \end{cases}$$

(3) $$\max\{x,y\} := \begin{cases} x & \text{for } x \geq y \\ y & \text{for } x < y. \end{cases}$$

1.2. Graph Theory

An *undirected graph* G is a pair G = (V,E), where V is a finite set of *nodes* or *vertices* and E is a family of unordered pairs of elements of V called *edges*. A directed graph, or *digraph*, is a graph with directions assigned to its edges. Formally, a digraph G is a pair G = (V,A) where V again is a set of vertices and A is a set of ordered pairs of vertices called *arcs*. Clearly, each directed graph gives rise to an underlying undirected graph by forgetting the orientation of the arcs. A (directed) graph with numerical values attached to its vertices and/or arcs is called a (directed) *network*.

In this book we are mostly concerned with directed networks. If not stated otherwise then n = #(V) and m = #(A) denote the number of vertices and arcs, respectively. For a digraph G = (V,A) and an arc a = (i,j) ϵ A we say that i is *adjacent* to j (and vice-versa) and that arc a is *incident* upon i and j. We say that arc (i,j) enters j and leaves i. The vertex i is called the *tail* and vertex j is called the *head* of arc (i,j). An arc (i,i) is called a *loop*. The *degree* of a vertex j of G is the number of arcs incident upon j. $\delta+(j)$ and $\delta-(j)$ denote the set of arcs leaving respectively entering vertex j. The head of an arc contained in $\delta+(j)$ is called a *successor* and the tail of an arc contained in $\delta-(j)$ a *predecessor* of j.

A graph is called *antisymmetric* if (i,j) ϵ A implies that (j,i) is not in A. A graph G' = (V',A') is a *subgraph* of G = (V,A) if $V' \subset V$ and $A' \subset A$. The graph G' is a *spanning subgraph* of G if V' = V.

A *path* P = P[i,k] from vertex i to vertex k is a list of vertices $[i = i_1, i_2, \ldots, i_q = k]$ such that $(i_p, i_{p+1}) \epsilon A$ or $(i_{p+1}, i_p) \epsilon A$ for all p = 1,2,...,q-1. Therein, all arcs of the form (i_p, i_{p+1}) are called *forward arcs*, and the arcs of the form (i_{p+1}, i_p) are called *backward arcs*. The set of forward and backward arcs is abbreviated by P^+ and P^-, respectively. The *characteristic vector* char(P) of a path P defined on A has components

$$\text{char}_{ij}(P) = \begin{cases} 1 & \text{if } (i,j) \ \epsilon \ P^+ \\ -1 & \text{if } (i,j) \ \epsilon \ P^- \\ 0 & \text{if } (i,j) \text{ is not contained in P.} \end{cases}$$

A path containing forward arcs only is called a *directed path*. A path is called *elementary* if there are no repeated vertices in its sequence. The number of arcs contained in a path P is denoted by #(P).

A *cycle* or *circuit* is a closed path $[i = i_1, i_2, \ldots, i_q = k]$ with $i = k$ and $q > 1$ without repeated vertices except for its starting and end point. A *directed cycle* is a closed directed path. For a connected subset $X \subset V$ a *cocycle* or *cut* (X, X^*) with $X^* := V - X$ is a subset of arcs $a = (i,j)$, where $i \in X$, $j \in X^*$ or $i \in X^*$, $j \in X$.

A graph is *connected* if there is a path between any two vertices in it. If not stated otherwise, we always consider connected graphs. A *tree* is a connected (sub) graph without cycles. We usually say that a tree $T = (V, A_T)$ spans its set of vertices V or is a *spanning tree*. Let T be a spanning tree of G. Then for each arc $a = (i,j) \in A - A_T$, there is a unique cycle $\mu(T,a)$ containing arcs from $A_T + \{a\}$ only. $\mu(T,a)$ is called *fundamental cycle* determined by T and a. The orientation of $\mu(T,a)$ is fixed by the orientation of a. The sets $\mu^+(T,a)$ and $\mu^-(T,a)$ summarize all the arcs which are directed according and opposite to the direction of $\mu(T,a)$, respectively. The family of all these cycles $\mu(T,a)$ for all $a \in A - A_T$ is called *fundamental cycle basis*. Similarly, for each $a \in T$, there is a unique cocycle denoted $\Omega(T,a)$ having arcs from $A - A_T + \{a\}$ only. $\Omega(T,a)$ is called *fundamental cocycle*, and the family of all these cocycles is called *fundamental cocycle basis*. Each cocycle can also be described by a connected subset X of V. We will use $\Omega^+(T,a) = \delta^+(X)$ for the set of arcs directed from X to X^*, and correspondingly $\Omega^-(T,a) = \delta^-(X)$ for the arcs directed from X^* to X.

A graph $G = (V,A)$ is called *bipartite* if V can be partitioned into two classes V_1 and V_2 such that each edge of A contains a vertex in V_1 and a vertex in V_2. It is easy to see that a graph is bipartite if and only if it contains no cycles of odd number of arcs. A graph is *planar* if it can be drawn on the plane such that no two edges (or arcs) cross one another.

The (vertex-arc) *incidence matrix* I(G) of a directed graph G is defined as $R = I(G)$ with

$$R(i,h) = \begin{cases} -1 & \text{if } i \text{ is the tail of arc } h \\ 1 & \text{if } i \text{ is the head of arc } h \\ 0 & \text{otherwise.} \end{cases}$$

There are different possibilities to represent a graph in a

computer. Besides the vertex-arc incidence matrix, the quadratic vertex-vertex adjacency matrix with (i,j)-th element equal to one if $(i,j) \in A$ and zero otherwise can be used. Another way is to store for each arc the tail and head vertex in an array. Finally, the most efficient ways to store a graph are the so-called *forward star* and *reverse star representation*. We only describe the forward star representation, the reverse star representation is analogous. For given numbers of vertices, the arcs of the graph must be numbered by a function $\tau: A \longmapsto \mathbf{N}$ such that for any two vertices $i,j \in V$ with $i < j$ and arcs (i,k) and (j,l) the corresponding numbers $\tau(i,k)$, $\tau(j,l)$ fulfill $\tau(i,k) < \tau(j,l)$. Additionally, a pointer point(i) is needed for each vertex i. The pointer gives the smallest number in the arc list of an arc leaving vertex i. The set of outgoing arcs of vertex i is stored from position point(i) to point(i+1) - 1 in the arc list. If point(i) = point(i+1) - 1 then vertex i has no outgoing arc. As a convention, we assign point(1) = 1 and point(n+1) = m + 1. See Figure 1.1. for an example.

Any algorithm requires one or more data structures to represent the elements of the problem to be solved and the information computed during the solution process. We will use lists, stacks and queues. A list is a sequence of elements which can be modified anywhere. The first element of a list is its head, the last element its tail. A *queue* is a list where insertions are restricted to the tail of the sequence and deletions are restricted to the head. For *stacks*, insertions and deletions are restricted to the head of the list.

Many graph algorithms require a mechanism for exploring the vertices and edges of a graph. The adjacency sets allow repeatedly to pass from a vertex to one of its neighbors. In general, there are many candidates for vertices visited next. Two techniques have been proven to be especially useful in exploring a graph. They are *breadth-first search* (BFS) and *depth-first search* (DFS) and result from the general search-procedure SEARCH by maintaining the list as a queue (BFS) and as a stack (DFS), respectively. At every stage within this procedure, each vertex j is either marked (label(j) = 1) or unmarked (label(j) = 0). In BFS the vertices are marked in nondecreasing order of the distance from the source 1, i.e., the minimum number of arcs for connecting the vertices with 1. BFS creates a path as long as possible, and backs up one vertex to initiate a new probe when it can mark no further vertices from the top of the path. Both search techniques are illustrated in Figure 1.2.

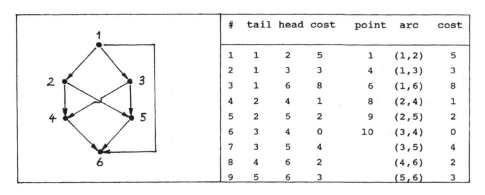

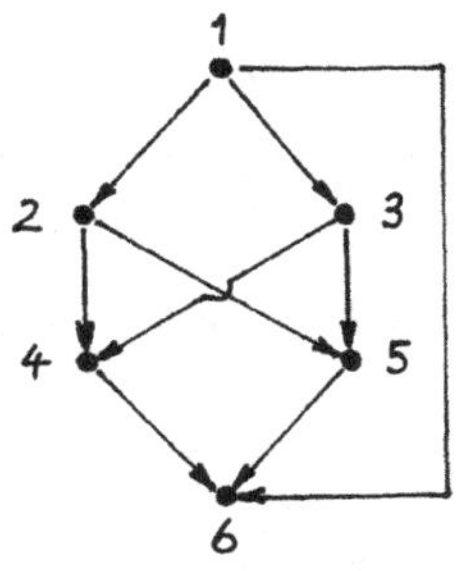

#	tail	head	cost	point	arc	cost
1	1	2	5	1	(1,2)	5
2	1	3	3	4	(1,3)	3
3	1	6	8	6	(1,6)	8
4	2	4	1	8	(2,4)	1
5	2	5	2	9	(2,5)	2
6	3	4	0	10	(3,4)	0
7	3	5	4		(3,5)	4
8	4	6	2		(4,6)	2
9	5	6	3		(5,6)	3

(a) (b) (c)

Figure 1.1. (a) Network.
(b) Arc list representation.
(c) Forward star representation.

```
procedure SEARCH
begin
  for all j ∈ V do label(j) := 0
  list := {1}
  label(1) := 1
  while list ≠ φ do
  begin
    choose i ∈ list
    for all j: (i,j) ∈ A do
    if label(j) = 0 then
    begin
      label(j) := 1
      list := list + {j}
    end
    list := list - {i}
  end
end
```

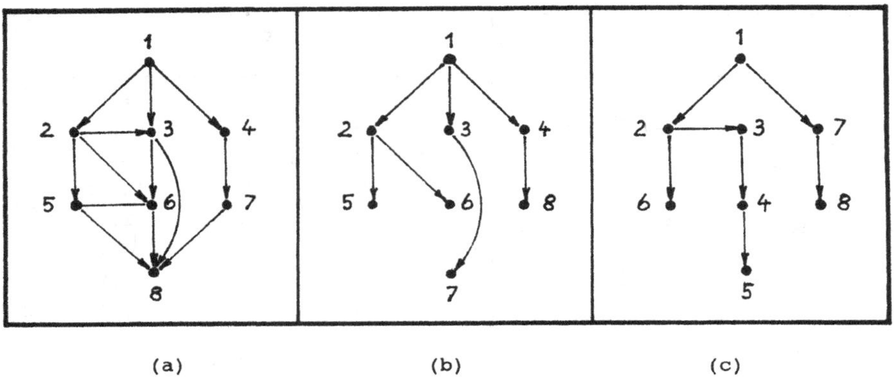

Figure 1.2. (a) Directed graph.
(b) Vertices as explored by BFS.
(c) Vertices as explored by DFS.

1.3. Algorithms and Complexity

In order to study algorithms and their complexity, we need a model of computation. Historically, the first such model proposed was the Turing machine (Turing 1936). In its simplest form a Turing machine consists of a finite state control, a two-way infinite memory tape divided into squares, each of which can hold one of a finite number of symbols, and a read/write head. In one step the machine can read the contents of one tape square, write a new symbol in the square, move the head one square left or right, and change the state of the control. Other computer models are the random-access machine or the pointer machine. Each of these models can simulate any other one with a polynomial loss in efficiency.

The *state* of an algorithm consists of the current values of all variables and the location of the current instruction to be executed. In a *deterministic algorithm*, each state upon execution uniquely determines at most one next state. A *nondeterministic algorithm* is one for which a state may determine many next states simultaneously. We may regard a nondeterministic algorithm as having the capability of branching off into many copies of itself, one for each next state.

Following Reingold, Nievergelt & Deo (1977), three special instructions are used in writing nondeterministic algorithms:

(i) x := *choice*(S) creates #(S) copies of the algorithm, and assigns every member of the set S to the variable x in one of the copies

(ii) *failure* causes that copy of the algorithm to stop execution.

(iii) *success* causes all copies of the algorithm to stop execution and indicates a "yes" answer to that instance of the problem.

The *running time* of an algorithm is the number of elementary operations used. By an *elementary operation* we mean addition, subtraction, multiplication, division, and comparison.

The *length* of an input is the number of bits occurring in it. For network problems as considered here, the input length is a polynomial function of n, m, log C and log U, where C und U are upper bounds on the cost coefficients respectively capacities associated with the set of arcs.

In studying the complexity of an algorithm, we are often interested in the behavior of the algorithm for very large inputs, only. Let $f(n), g(n): \mathbf{N} \longmapsto \mathbf{R}$. We write $f(n) = O(g(n))$ if there exists a constant $c > 0$ such that, for large enough n, $f(n) \le c \cdot g(n)$.

There are different approaches for analyzing the performance of an algorithm: worst case analysis, average case analysis, and empirical analysis. The *worst-case complexity* of an algorithm $\mathcal{L}$ is the smallest function f such that $\mathcal{L}$ runs in O(f(n)) time. It provides a performance guarantee; the algorithm will always require no more time (or space, if storage space is taken into account) than specified by the bound. The disadvantage of this measure are pathological cases determining the complexity even though they may be exceedingly rare in practice. In the average-case analysis, the average number of steps of an algorithm, taken over all inputs of a fixed size n, is counted. To do this, a probability distribution for the instances must be chosen. To find an adequate one and handling it mathematically, especially when independence cannot be assumed, is a difficult task.

The objective of empirical analysis is to measure the computational time of an algorithm on a series of randomly generated test examples. This implies an estimation how algorithms behave in practice.

The *complexity of a problem* Π is the smallest function f for which there exists an O(f(n))-time algorithm for Π, i.e., the minimum complexity over all possible algorithms solving Π. An (deterministic) algorithm runs in *polynomial time* if the number of operations is bounded by a polynomial in the input length. In the context of networks, an algorithm is polynomial if its running time is polynomial in n, m, log C, and log U. A polynomial algorithm is said to be *strongly polynomial* if its running time is bounded by n and m only, and does not include terms in log C and log U. Finally, an (network) algorithm is said to be *pseudopolynomial* if its running time is polynomial in n, m, C, and U.

A problem Π is in the class P if there exists a (deterministic) polynomial-time algorithm which solves Π. A problem Π is in the class NP (nondeterministic polynomial) if there exists a nondeterministic polynomial-time algorithm which solves Π. One problem Π_1 is *polynomially reducible* to another problem Π_2, written $\Pi_1 \propto \Pi_2$, if there exists a (deterministic) polynomial-time function f mapping the instances of Π_1 into instances of Π_2 such that for all instances I a solution f(I) of Π_2 gives a solution of the original instance.

A problem Π is called *NP-hard* if $\Pi' \propto \Pi$ for all $\Pi' \in$ NP. A problem Π is called *NP-complete* if it is NP-hard and in NP. Thus, every NP-complete problem Π has the property that if Π can be solved in (deterministic) polynomial time than all NP-problems can be solved in (deterministic) polynomial time. For that reason, NP-complete problems are considered to be the "hardest" problems within NP. For the first member of this class Cook (1970) proved that the 'Satisfiability' problem of mathematical logic is NP-complete. Now, hundreds of problems from different areas are proven to be in this class.

§2 MAXIMUM FLOWS

2.1. Problem Statement and Fundamental Results

The maximum flow problem is one of the fundamental problems in combinatorial optimization. The problem has numerous applications, especially in the area of transportation and communication networks. Moreover, there are a variety of problem solving procedures using the calculation of maximum flows as a subroutine.

We consider a connected, antisymmetric graph $G = (V,A)$ without loops. We assume two special vertices: the source 1 and the sink n with no entering and no leaving arcs, respectively. Without loss of generality we exclude parallel arcs since their occurrence can always be circumvented by introducing dummy arcs and vertices. With the number of arcs equal to m, a vector $x \in \mathbf{R}^m$ defined on A is called a *flow* if

$$(1) \quad \Sigma_{(j,k)\in A}\, x(j,k) - \Sigma_{(i,j)\in A}\, x(i,j) = \begin{cases} \#(x) & \text{for } j = 1 \\ -\#(x) & \text{for } j = n \\ 0 & \text{otherwise} \end{cases}$$

where $\#(x) := \Sigma_{(1,k)\in A}\, x(1,k)$ is called the *value* of the flow. A flow is said to be *maximum* if it is of maximum value. With each arc $(i,j) \in A$ a positive capacity $cap(i,j)$ is associated . A flow is called *feasible* if

$$(2) \quad 0 \le x(i,j) \le cap(i,j) \quad \text{for all } (i,j) \in A.$$

The constraints (1),(2) define the *flow polyhedron* $\mathbf{X}$. The maximum flow problem is

MF $\max \{\#(x)\colon x \in \mathbf{X}\}$.

Using $X^* := V - X$, a *separating cut* (X,X^*) is a cut such that $1 \in X$ and $n \in X^*$. If not stated otherwise, $\delta^+(X)$ is identified with the set of arcs $(i,j) \in A\colon i \in X,\ j \in X^*$. Correspondingly, $(i,j) \in \delta^-(X)$ means that $(i,j) \in A$ with $i \in X^*$ and $j \in X$. The set of all these cuts is abbreviated by C. The capacity $Cap(X,X^*)$ of a cut is defined as

$$(3) \quad Cap(X,X^*) := \Sigma_{(i,j)\in\delta+(X)}\, cap(i,j).$$

A cut of minimum capacity is called a *minimum cut*; the set of all minimum cuts is denoted by C_{min}. Correspondingly, X_{max} abbreviates the set of all maximum flows.

Theorem 2.1. (Ford & Fulkerson 1962)
A flow $x \in \mathbf{X}$ is maximum if and only if there is a separating cut (X,X^*) such that $\#(x) = Cap(X,X^*)$.

■

A matrix B is *totally unimodular* if each subdeterminant of A is 0,+1, or -1. In particular, each entry in a totally unimodular matrix must be 0,+1, or -1. From the total unimodularity of the vertex-arc incidence matrix I(G) we obtain as a special case of a result of Hoffman & Kruskal (1956):

Theorem 2.2.
If all capacities in MF are integers, there is a maximum flow with integers on every arc.

■

2.2. Augmenting Paths and Blocking Flows

Many algorithms for the maximum flow problem are based on the concept of flow augmenting paths. We consider a feasible flow $x \in \mathbf{X}$ and define the *residual graph* $R(x) = (V,A(x))$ with vertex set V and

$$A(x) := A^+(x) + A^-(x) \text{ where}$$
$$A^+(x) := \{(i,j): (i,j) \in A \;\&\; x(i,j) < cap(i,j)\},$$
$$A^-(x) := \{(i,j): (j,i) \in A \;\&\; x((j,i) > 0\}.$$

The residual capacity res: $A(x) \longmapsto \mathbf{R}$ with respect to x is

$$(4)\quad res(i,j) := \begin{cases} cap(i,j) - x(i,j) & \text{for } (i,j) \in A^+(x) \\ x(i,j) & \text{for } (i,j) \in A^-(x). \end{cases}$$

A *flow augmenting path* related to x is a directed path P in R(x) from 1 to n.

Lemma 2.1. (Ford & Fulkerson 1962)
A flow x is maximum if and only if there is no flow augmenting path in R(x).

Proof:

If there is a flow augmenting path P in R(x), then we can push up to res(P) := min { res(i,j): (i,j) ϵ P } > 0 units from 1 to n. This leads to a new flow x^* with $\#(x^*) := \#(x) + res(P) > \#(x)$. Consequently, the 'only if'-part is valid. For the 'if'-part we consider the subset $X \subset V$ which is reachable from 1 in R(x). X must be a proper subset of V, otherwise there would be a flow augmenting path. For the cut (X,X^*) it holds

$$\begin{aligned} Cap(X,X^*) &= \Sigma_{(i,j)\epsilon\delta+(X)}\ cap(i,j) \\ &= \Sigma_{(i,j)\epsilon\delta+(X)}\ x(i,j) \quad = \#(x). \end{aligned}$$

From Theorem 2.1. it follows that x is maximum.

■

Lemma 2.1. is the base of the augmenting path method developed by Ford & Fulkerson (1962) which is summarized in procedure AUGMENT.

```
procedure AUGMENT
begin
  for all (i,j) ε A do x(i,j) := 0
  #(x) := 0
  compute R(x)
  while n is reachable from 1 in R(x) do
  begin
    calculate a flow augmenting path P in R(x)
    res(P) := min {res(i,j): (i,j) ε P}
    for all (i,j) ε P  A+(x) do x(i,j) := x(i,j) + res(P)
    for all (i,j) ε P  A-(x) do x(j,i) := x(j,i) - res(P)
    update R(x)
    #(x) := #(x) + res(P)
  end
end
```

As the path P in the loop can be chosen arbitrarily, the above procedure needs $O(m \cdot \#(x))$ steps for integer capacities which is a bad performance since it depends on #(x) directly. Moreover, in the case of irrational capacities the method may fail in the sense that it may need an infinite number of iterations. Additionally, the sequence of flows need not converge to a maximum flow. However, the augmenting path method works well in practice. The main reason is that the method is mostly implemented using augmenting paths of minimum number of arcs. With this additional property, Edmonds & Karp (1972) proved a total running time of $O(n \cdot m^2)$.

Dinic (1970) introduced the concept of a blocking flow and suggested an algorithm that augments in each iteration along all paths of shortest length in R(x). To define the *layered graph* L(x) with respect to x we need the shortest distance (number of arcs) dist(j) in R(x) from 1 to j for each $j \in V$. If j is not reachable in R(x) we define dist(j) := ∞. Then L(x) = (V,AL(x)) is a subgraph of R(x) with

(5) $(i,j) \in AL(x)$ if $(i,j) \in A(x)$ and dist(j) = dist(i) + 1

Each k: $0 \leq k \leq$ dist(n) defines a layer in the network with set of vertices V(k) := $\{j \in V$: dist(j) = k$\}$. A flow $x \in \mathbf{X}$ is called *blocking* if each path from 1 to n in L(x) contains an arc $(i,j) \in A(x)$ with res(i,j) = 0. The layered graph can be computed from the residual graph in O(m) steps by means of breadth-first search. The following procedure LAYER uses a subroutine BLOCKING to calculate a blocking flow dx with respect to the current flow x.

```
procedure LAYER(cap)
begin
  for all (i,j) ∈ A  do x(i,j) := 0
  compute L(x)
  while n is reachable from 1 in L(x) do
  begin
    dx := BLOCKING(x)
    for all (i,j) ∈ A⁺(x) ∩ AL(x) do x(i,j) := x(i,j) + dx(i,j)
    for all (i,j) ∈ A⁻(x) ∩ AL(x) do x(j,i) := x(j,i) - dx(i,j)
    update L(x)
  end
end
```

We illustrate LAYER by the example given in Figure 2.1. There are two blocking flows in layered graphs having an increasing number of layers. We remark that in the final graph, both vertices 4 and 6 have distance equal to ∞. The total maximum flow results from summarizing all the calculated blocking flows. The value of the maximum flow is equal to six.

Lemma 2.2.
Procedure LAYER needs the calculation of O(n) blocking flows to determine an optimal solution of MF.

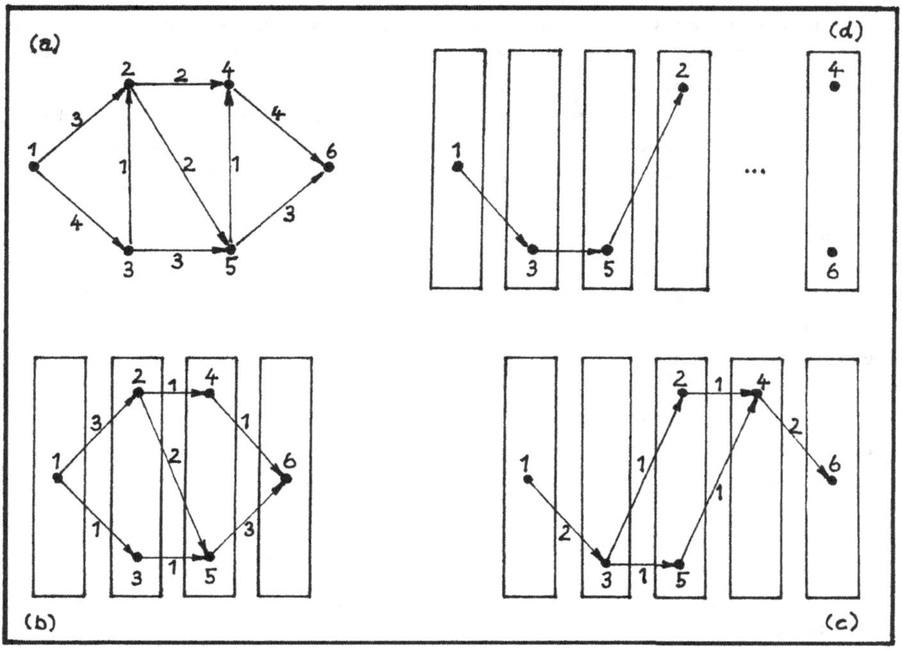

Figure 2.1. (a) Graph G with capacities cap(i,j).
(b) Blocking flow dx in $L(x^0)$ for $x^0 := 0$.
(c) Blocking flow dx^1 in $L(x^1)$ for $x^1 := x^0 + dx^1$.
(d) Layered graph $L(x^2)$ for $x^2 := x^1 + dx^2$.

Proof:
As a consequence of Lemma 2.1. we obtain that for $x \in \mathbf{X}$ and dist(n) = ∞ in L(x) it follows the optimality of x. Assume dist(n) < ∞. Let dx be a blocking flow in L(x). Since dx summarizes all augmenting paths of length dist(n), it follows that dist(n) is increased by 1. Using $1 \leq \text{dist}(n) \leq n-1$ results in the stated complexity bound. Assume, the calculated flow x is not maximum. The existence of a flow augmenting path in R(x) follows from Lemma 2.1. Consequently, dist(n) < ∞ and the algorithm does not stop with x.

■

From the definition of L(x) it follows that the layered graph is acyclic, i.e., it does not contain a directed cycle. So it remains to calculate iteratively blocking flows in acyclic graphs. This can be done in different ways. Dinic (1970) applied depth-first search to find a path P from 1 to n in L(x). Pushing enough flow along P results in a saturation for at least one arc. After updating L(x) this process is repeated until there is no path in L(x). A more precise description is given below as procedure DINIC. The complexity of this algorithm is $O(n \cdot m)$ to find a blocking flow and $O(n^2 \cdot m)$ to find a maximum flow.

```
procedure DINIC
begin
C Calculation of a blocking flow dx in L(x) = (V,AL(x))
  for all (i,j) ∈ AL(x) do
  begin
    dx(i,j) := 0
    label(i,j) := 'unblocked'
  end
  repeat
    j := 1
    S := ∅ ('S is a stack')
    repeat
      if there is no (j,k) ∈ AL(x): label(j,k) := 'unblocked' then
      begin
        if j := 1 then stop ('the present flow dx is maximal')
        S := S - {(i,j)} ('(i,j) is the top-most element of S')
        label(i,j) := 'blocked'
        j := i
      end
      S := S + {(j,k)} for an unblocked (j,k) ∈ AL(x)
      j := k
    until j := n
C The edges on S form a flow augmenting path
    δ := min { res(i,j): (i,j) ∈ S }
    for all (i,j) ∈ S do
    begin
      dx(i,j) := dx(i,j) + δ
      res(i,j) := res(i,j) - δ
      if res(i,j) := 0 then label(i,j) := 'blocked'
    end
  until n is not reachable in L(x)
end
```

Further progress in worst-case complexity was achieved by Karzanov

(1974) using the concept of preflows which is considered in Section 2.4. in more detail. The approach using these preflows needs $O(n^2)$ time to calculate a blocking flow in L(x). The overall complexity is $O(n^3)$. Sleator & Tarjan (1983) implemented Dinic's algorithm using a clever data structure called dynamic trees in $O(m \cdot n \cdot \log n)$ time. Goldberg (1985) used the idea of preflows, but applied it to the entire residual network rather than the level network. Instead of storing levels exactly, a lower bound on the distance from each vertex to the sink is taken. These distance labels are a better computational device than layered networks because of the distance labels are simpler to understand and easier to manipulate. An important feature of this approach is the fact that all computations are locally computable. This is of great importance from the point of view of parallel computing. Using dynamic trees again results in a speed up of the bottleneck operation (unsaturated pushes) of Goldberg' algorithm and a running time of $O(n \cdot m \cdot \log(n^2/m))$ due to Goldberg & Tarjan (1986). Ahuja & Orlin (1987) developed a modification of Goldberg's algorithm in which the bottleneck operation of non-saturating pushes can be reduced to $O(n^2 \cdot \log U)$ by using excess scaling. Here and in the following, U denotes an upper bound on the integral arc capacity. Capacity scaling in conjunction with the maximum flow problem is discussed in Section 2.3. The scaling algorithm for the maximum flow problem can be further improved by using more clever rules to push flow or by using dynamic trees. These results are described in Ahuja, Orlin & Tarjan (1988).

In Table 2.1., a history of maximum flow algorithms with improvements in the worst-case complexity (at least for some classes of problems) is given. The algorithms signed with (*) are based on layered graphs and differ only in the way the blocking flows are calculated.

Table 2.1. History of the worst-case complexity of maximum flow algorithms.

Author		Complexity
Ford & Fulkerson (1956)		unbounded for real numbers $O(m\ \#(x))$ for rationals
Dinic (1970)	(*)	$O(n^2 \cdot m)$
Edmonds & Karp (1972)		$O(n \cdot m^2)$
Karzanov (1974)	(*)	$O(n^3)$
Cherkassky (1977)	(*)	$O(n^2 \cdot m^{1/2})$
Malhotra et al.(1978)		$O(n^3)$
Shiloach (1978)	(*)	$O(n \cdot m(\log n)^2)$
Galil (1980)	(*)	$O(n^{5/3} \cdot m^{2/3})$
Galil & Naamad (1980)	(*)	$O(n \cdot m(\log n)^2)$
Sleator & Tarjan (1983)	(*)	$O(n \cdot m \cdot \log n)$
Gabow (1985)	(*)	$O(n \cdot m \cdot \log U)$
Goldberg (1985)		$O(n^3)$
Goldberg & Tarjan (1986)		$O(n \cdot m \cdot \log(n^2/m))$
Ahuja & Orlin (1986)		$O(n \cdot m + n^2 \cdot \log U)$
Ahuja & Orlin (1988)		$O(n \cdot m + (n^2 \cdot \log U)/\log\log U)$
Ahuja, Orlin & Tarjan (1988)		$O(n \cdot m + n^2 \cdot (\log U)^{1/2})$

2.3. Scaling

Scaling was introduced by Edmonds & Karp (1972) to solve the minimum-cost flow problem and later extended by Gabow (1985) for other network optimization problems. In case of MF, scaling iteratively transforms the given problem into another one with scaled capacities, i.e., a number k becomes $\lfloor k/2 \rfloor$. The solution of the scaled down problem is doubled and forms a near-optimum solution of the previous problem. It remains to transform the near-optimum to an optimum solution. This idea was used by Gabow (1985).

Lemma 2.3.
Let x^* be an optimal solution of the maximum flow problem with scaled capacities $cap^*(i,j) := \lfloor (cap(i,j)/2) \rfloor$, then

(i) $\#(x1) \le m + 2 \cdot \#(x^*)$ for any optimal flow x1 of the original problem MF, and

(ii) in $G(2 \cdot x^*)$ there are at most m flow augmenting paths with residual capacity not greater than 1.

Proof:

$$\begin{aligned} & \max \{ \#(x) : x \in \mathbf{X} \} \\ = & \#(x1) \\ = & \min \{ Cap(X,X^*) : (X,X^*) \in C \} \\ \le & \min \{ \Sigma_{(i,j) \in \delta+(X)} (2 \cdot \lfloor (cap(i,j)/2) + 1 \rfloor : (X,X^*) \in C \} \\ \le & m + 2 \cdot \min \{ \Sigma_{(i,j) \in \delta+(X)} cap^*(i,j) : (X,X^*) \in C \} \\ = & m + 2 \cdot \#(x^*). \end{aligned}$$

Since x^* is an optimal solution for cap^* it follows that in the residual graph $G(2 \cdot x^*)$ there cannot be a flow augmenting path with residual capacity greater than 1. As a consequence of $\#(x1) - 2 \cdot \#(x^*) \le m$, the number of such paths is bounded by m.

■

We give a formal description of the scaling approach applied to MF. Procedure SCALE is a recursive algorithm and uses LAYER as a subprogram.

```
procedure SCALE(x,cap)
begin
  while max { cap(i,j): (i,j) ∈ A } > 1 do
  begin
    for all (i,j) ∈ A do scap(i,j) := ⌊(cap(i,j)/2)⌋
    x := SCALE(x,scap)
    compute L(x)
    for all (i,j) ∈ A do rcap(i,j) := cap(i,j) - 2·x(i,j)
    dx := LAYER(rcap)
    x := 2x + dx
  end
end
```

Lemma 2.4.
Procedure SCALE needs $O(n \cdot m \cdot \log U)$ steps to solve MF.

Proof:
The number of recursive calls is bounded by log U. The solution of each subproblem to compute an optimal solution x1 from $2 \cdot x^*$ needs O(n) blocking flows. Each phase h needs $O(m + r_h \cdot n)$ steps, where r_h is the number of augmenting paths in phase h. From $\Sigma\, r_h = m$ the given bound is obtained.

■

The procedure SCALE is illustrated by the numerical example presented by the figures below. Table 2.2. summarizes the numerical values occurring in the iterations of the algorithm.

Table 2.2. Numerical values occurring during the algorithm with $cap^k(i,j) := \lfloor (cap^{k-1}(i,j)/2) \rfloor$ for all $(i,j) \in A$ and $k = 1,2,3$.

(i,j)	(1,2)	(1,3)	(1,4)	(2,5)	(3,7)	(4,6)	(4,7)	(5,3)	(5,8)	(6,2)	(6,8)	(7,8)
cap^0	3	2	9	6	5	7	1	7	2	2	4	6
cap^1	1	1	4	3	2	3	0	3	1	1	2	3
cap^2	0	0	2	1	1	1	0	1	0	0	1	1
cap^3	0	0	1	0	0	0	0	0	0	0	0	0
x^0	3	2	7	5	5	6	1	3	2	2	4	6
x^1	1	1	3	2	2	3	0	1	1	1	2	2
x^2	0	0	1	0	0	1	0	0	0	0	1	0
x^3	0	0	0	0	0	0	0	0	0	0	0	0

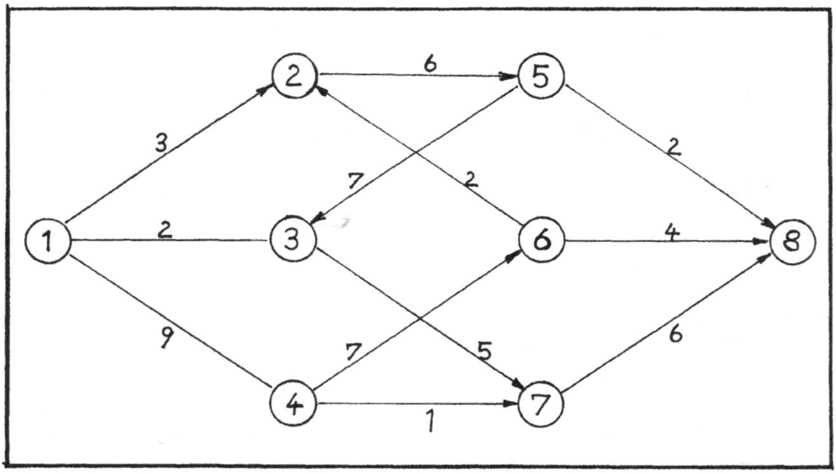

Figure 2.2. Graph G = (V,A) with capacities cap(i,j).

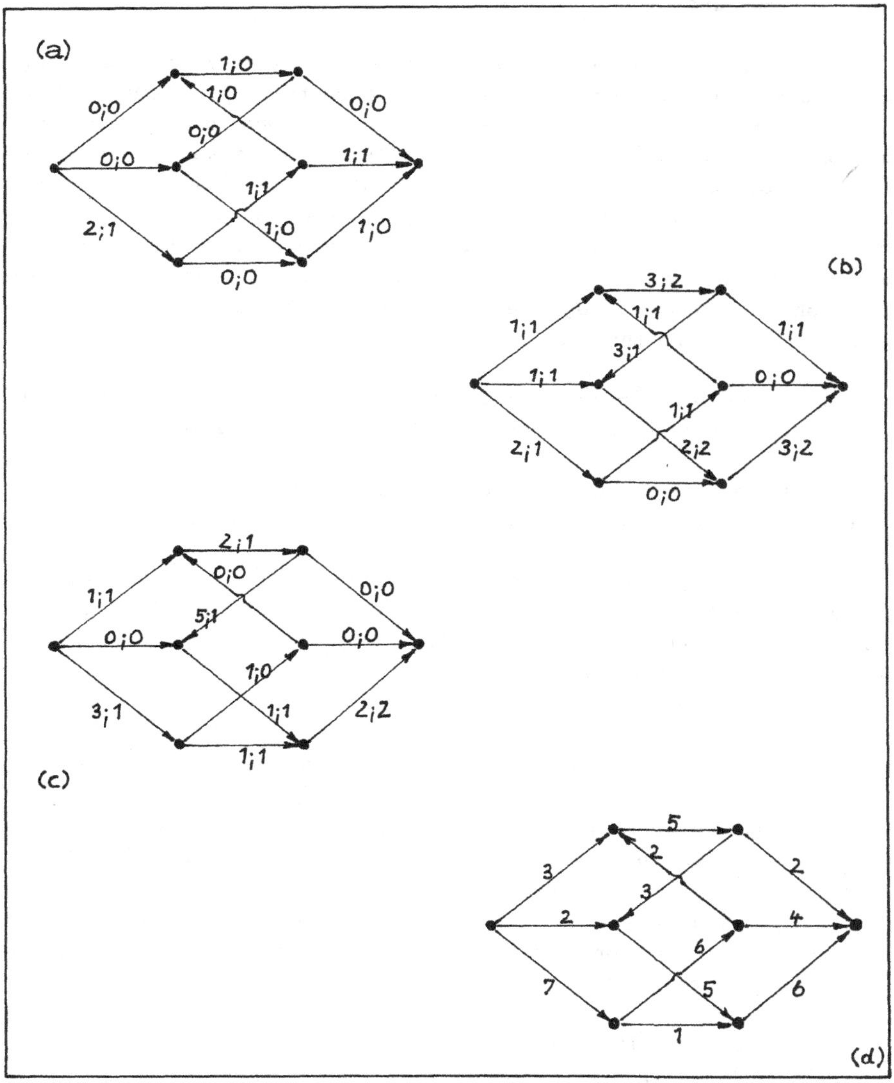

Figure 2.3. Graph G with values rcap(i,j);dx(i,j).
(a) $rcap^2 = cap^2 - 2x^3$ and max flow dx based on $rcap^2$;
(b) $rcap^1 = cap^1 - 2x^2$ and max flow dx based on $rcap^1$;
(c) $rcap^0 = cap^0 - 2x^1$ and max flow dx based on $rcap^0$;
(d) maximum flow x^0.

2.4. Preflows and the Goldberg Algorithm

The blocking flow approach of Dinic firstly determines a candidate for a maximum flow x until vertex n is not contained in L(x). In that case the set of reachable vertices and its complement define a separating cut of capacity equal to the maximum flow value. The preflow algorithm of Goldberg (1985) firstly computes a candidate for a minimum cut and subsequently determines a flow whose value is equal to the capacity of that cut. The algorithm maintains a preflow in the original network and pushes flow towards the sink along what is estimated to be shortest paths. Only when the algorithm terminates the preflow becomes a flow, and then it is a maximum flow.

The notion of a preflow was introduced by Karzanov (1974) for computing blocking flows. A vector $g: V \longmapsto \mathbf{R}$ is called a *preflow* if

(i) $0 \le g(i,j) \le cap(i,j)$ for all $(i,j) \in A$, and
(ii) $\Sigma_{(j,k)\in A}\, g(j,k) - \Sigma_{(i,j)\in A}\, g(i,j) \le 0$ for all $j \in V - \{1\}$.

For a given preflow g, the residual graph R(g) and the residual capacity are introduced in analogy to flows by replacing the flow variables by preflow variables. Furthermore, for each vertex $j \in V$ we define the excess function

$$(6)\quad e(j) := \begin{cases} \infty & \text{for } j = 1 \\ \Sigma_{(i,j)\in A}\, g(i,j) - \Sigma_{(j,k)\in A}\, g(j,k) & \text{otherwise.} \end{cases}$$

Clearly, a preflow g is a flow if $e(j) = 0$ for all vertices $j \in V - \{1,n\}$. A fundamental role in the new algorithm plays a distance label d(j) which is an estimate of the length of the shortest augmenting path from j to n. Let g be a preflow. A non-negative integer-valued function $d: V \longmapsto \mathbf{N}$ is called a *distance function* if the following properties are fulfilled:

(i) $d(n) = 0$, $d(j) > 0$ for all $j \ne n$, and
(ii) $d(i) \le d(j) + 1$ for each $(i,j) \in R(g)$ such that $d(i) < n$ or $d(j) < n$.

Lemma 2.5.
The function d gives a lower bound on the distance from j to n in R(g) for all vertices $j \in V$.

Proof:

Consider $j \in V$ and a path P from j to n in R(g) having k arcs ($\#(P) = k$). The proposition is shown by induction: $k = 0$ implies that $j = n$ and $d(j) = 0$. Let $P = [j = j_0, j_1, \ldots, j_k = n]$ and $P^* = [j_1, \ldots, j_k = n]$. Then $\#(P^*) = k-1$ and $d(j_1) \leq \#(P^*)$ due to our assumption for k-1. From $d(i) \leq d(j) + 1$ for all $(i,j) \in R(g)$ it follows that $\#(P) = \#(P^*) + 1 \geq d(j_1) + 1 \geq d(j)$.

■

As examples for distance functions we give

$$(7)\quad d^1(i) = \begin{cases} 0 & \text{for } i = n \\ 1 & \text{otherwise} \end{cases}$$

and

$$(8)\quad d^2(i) = \begin{cases} \min\{\#(P_i) : P_i \text{ is a path from } i \text{ to } n \text{ in } R(x)\} \\ n \quad \text{otherwise.} \end{cases}$$

Lemma 2.6.

Let g be a preflow and d be a distance function. If $d(j) \geq n$ for a vertex $j \in V$ then there is no path from j to n in R(g).

Proof:

Assume the existence of a path from j to n in R(x). Then there is also a shortest path P from j to n in R(g). Since a shortest path from a vertex j to n in R(g) is at most of length n-1 it follows that $n-1 \geq \#(P) \geq d(j)$ in contradiction to $d(j) \geq n$.

■

A vertex $j \in V$ is called *active* if $0 < d(j) < n$ and $e(j) > 0$. The algorithm of Goldberg contains two phases; the first one determines a minimum cut, and in the second one the resulting preflow is transformed into a (maximum) flow. In the formal description given below, it is assumed that for each vertex j there is maintained a list N(j) of adjacent vertices. Additionally, each vertex j has also a current neighbor cn(j) from this list.

```
procedure GOLDBERG I
begin
  g := 0
  computation of a valid labeling
  for all j ∊ V do cn(j) := first element of N(j)
  Q := {1}
  while Q ≠ ∅ do
  begin
    select any active vertex j ∊ Q
    Q := Q - {j}
    k := cn(j)
    while d(k) ≥ d(j)  or res(j,k) > 0 do
    begin
      if k = last element of N(j) then RELABEL
                                  else cn(j) := next element of N(j)
      k := j
    end
    PUSH
    if j is active then Q := Q + {j}
  end
end

PUSH
begin
  δ := min {e(j),res(j,k)}
  if (j,k) ∊ A then g(j,k) := g(j,k) + δ
  if (k,j) ∊ A then g(k,j) := g(k,j) - δ
  if k is active then Q := Q + {k}
end

RELABEL
begin
  cn(j) := first element of N(j)
  d(j) := min {d(k) + 1: (j,k) ∊ R(g)}
end
```

Let g be a preflow determined by procedure Goldberg I. Define $X^* = \{j \in V:$ there is a path from j to n in R(g)$\}$ and $X := V - X^*$.

Lemma 2.7. (Goldberg 1985)
Procedure Goldberg I calculates a preflow g and a cut (X,X^*) with the following properties:

(i) $(i,j) \in \delta^+(X)$ implies $g(i,j) = c(i,j)$

(ii) $(i,j) \in \delta^-(X)$ implies $g(i,j) = 0$

(iii) $e(j) = 0$ for all vertices $j \in X^*$, $j \neq n$.

If Q is maintained as a stack, the algorithm runs in $O(n^2 \cdot m)$ time. ■

In the second phase of the algorithm, the excess of vertices of X is returned to 1 along estimated shortest paths. Therefore, a *vertex labeling* dl: $dl(1) = 0$ and $dl(j) \leq dl(i) + 1$ for all edges $(i,j) \in R(g)$ is considered. Examples are

$$dl(j) = \begin{cases} 0 & \text{for } j = 1 \\ 1 & \text{otherwise} \end{cases}$$

or the level function dist from the layered graph approach. Due to definition, the last one is the best possible labeling.

A vertex $j \in V - \{n\}$ is called *violating* if

(i) $0 < dl(j) < n$

(ii) $e(j) > 0$.

The second phase consists of repeating the following steps in any order until there is no violating vertex.

procedure GOLDBERG II

REDUCE

```
begin
  select a violating vertex j ∈ X
  select an edge (j,k) ∈ R(g) with dl(k) = dl(j) + 1
  δ := min {e(j),g(j,k)}
  g(j,k) := g(j,k) - δ
end
```

RELABEL

```
begin
  select any vertex j : 0 < dl(j) < n
  dl(j) := min {dl(k) + 1 : (j,k) ∈ R(g)}
end
```

Theorem 2.3. (Goldberg 1985)
The second phase terminates after $O(n \cdot m)$ steps with a maximum flow g and proving (X,X^*) to be a minimum cut. The overall Goldberg algorithm needs $O(n^2 \cdot m)$ steps for stack respectively $O(n^3)$ steps for queue manipulation of Q.

■

Computational results of Derigs & Meier (1989) indicate that the above algorithm finds a minimum cut quite quickly. However, it needs a lot of time to verify this according to the termination condition of phase I. As a consequence, Derigs & Meier (1988) proposed a new termination rule based on the concept of a gap.

Let g be a preflow and d be a distance function. A positive integer β is called a *gap* if

(i) $\beta < n$
(ii) $d(j) \neq \beta$ for all $j \in V$
(iii) $\{j \in V: d(j) > \beta\} \neq \emptyset$.

With β a gap, a new distance function is used:

$$(9) \quad d(j) = \begin{cases} n & \text{if } d(j) > \beta \\ d(j) & \text{otherwise.} \end{cases}$$

Then the RELABEL step in procedure Goldberg I is modified as follows:

```
RELABEL-GAP
begin
  cn(j) := first element of N(j)
  d(j) := min {d(k) + 1 : (j,k) ∈ R(g)}
  if there is a gap β then
  for all j ∈ V  do if d(j) > β then d(j) := n
end
```

We illustrate procedure Goldberg by the network of Figure 2.2. A valid distance function is given by d = (3,2,2,2,1,1,1,0). The sets N(j) of adjacent vertices are:

N(1) = {2,3,4}	N(5) = {3,8,2}
N(2) = {5,6,1}	N(6) = {2,8,4}
N(3) = {7,5,1}	N(7) = {8,3,4}
N(4) = {6,7,1}	N(8) = {5,6,7}

PUSH: g(1,2) = 3
PUSH: g(2,5) = 3
PUSH: g(5,8) = 2

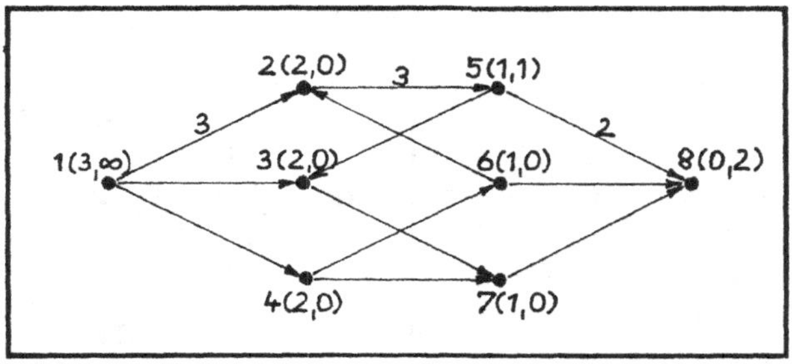

RELABEL : d(5) = 3
PUSH : g(5,3) = 1
PUSH : g(3,7) = 1
PUSH : g(7,8) = 1

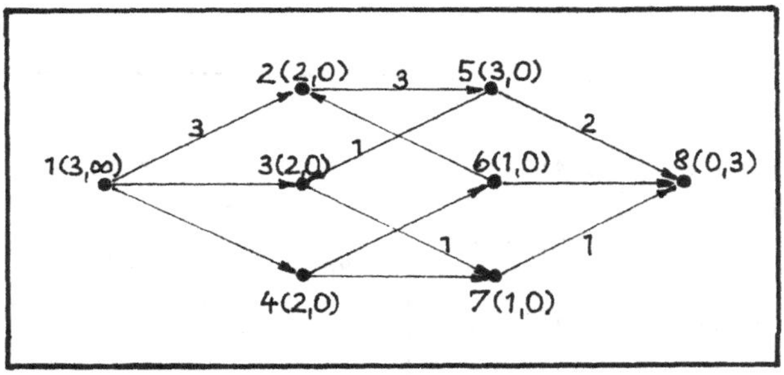

PUSH : g(1,3) = 2
PUSH : g(3,7) = 2
PUSH : g(7,8) = 2
PUSH : g(1,4) = 9
PUSH : g(4,6) = 7
PUSH : g(6,8) = 4

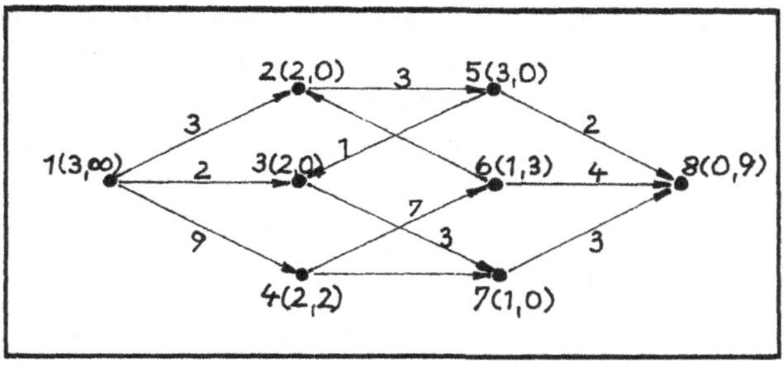

Figure 2.4. (a) Subsequent iterations of procedure GOLDBERG I. For each vertex i, the distance function d(i) and excess e(i) are given in the triple i(d(i),e(i)). The numbers along the arcs represent the current preflow.

RELABEL : d(6) = 3
PUSH : g(6,2) = 2
RELABEL : d(2) = 4
PUSH : g(2,5) = 5
PUSH : g(5,3) = 3
PUSH : g(3,7) = 5
PUSH : g(7,8) = 5

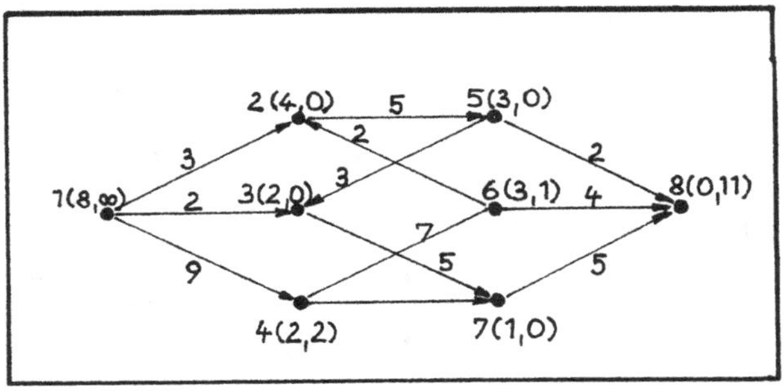

PUSH : g(6,4) = 1
PUSH : g(4,7) = 1
PUSH : g(7,8) = 1
RELABEL : d(4) = 4
RELABEL : d(6) = 5
RELABEL : d(4) = 6
RELABEL : d(6) = 7

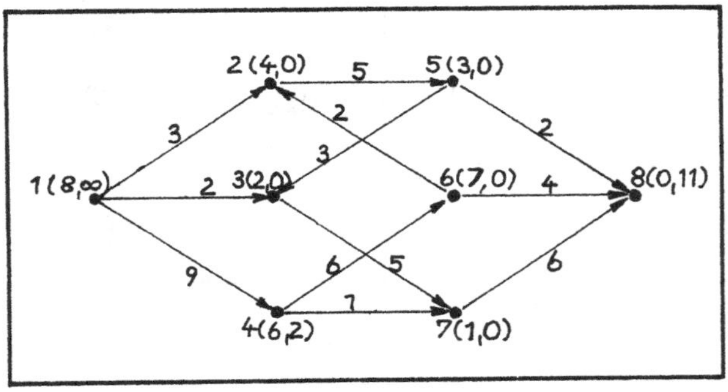

Figure 2.4.(b) Subsequent iterations of procedure GOLDBERG II. For each vertex i, the distance function d(i) and excess e(i) are given in the triple i(d(i),e(i)).The numbers along the arcs represent the current pre flows.

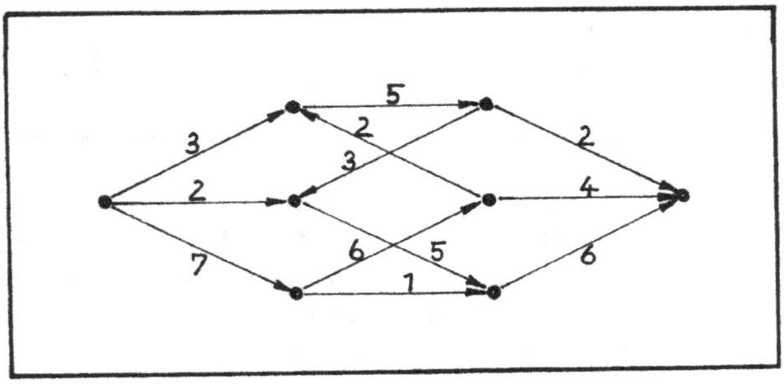

Figure 2.4.(c) Maximum flow at the end of GOLDBERG II.

At the beginning, the current neighbors are cn = (2,5,7,6,3,2,8,5). At the end of the first phase, with $\beta = 5$ a minimum cut with X = {1,4,6} is defined. In the second phase, the excess of the one violating vertex $4 \in X$ is returned to the source 1. With the level function we obtain dist(1) = 0, dist(4) = 1, and one REDUCE-operation is necessary to obtain the maximum flow of Figure 2.4.(c).

2.5. Computational Results

Comprehensive computational studies comparing various maximum flow algorithms have been made by Hamacher (1979), Cheung (1980), Glover et al. (1979) , Imai (1983) and Derigs & Meier (1989).

Derigs & Meier (1989) discussed several ideas for implementing and enhancing the approach of Goldberg (1985). They presented computational results for these implementations and compared the best Goldberg-code with implementations of the Dinic-method and the Karzanov-method. The different versions of the Goldberg-code resulted from a variation of:

(i) the choice of the distance function (naive labeling (7) or exact labeling (8) in phase 1 and the corresponding functions in phase 2) ,

(ii) the application of gap-search, and

(iii) the strategy to select the next active respectively violating vertex in both phases.

For (iii), the two classical concepts first-in-first-out selection (FIFO-rule) and last-in-first-out selection (LIFO-rule) were considered. The FIFO-rule was realized by a QUEUE while the LIFO-rule was implemented using a stack. Additionally, another data structure, the so-called DEQUEUE (doubly ended queue) was considered. In the last case, the following strategy was used:

- if a vertex becomes active or violating for the first time it is added at the top of the STACK, and
- if a vertex that has been active before, becomes active again it is added at the end of the QUEUE.

As an alternative to the above list-search techniques, in the

- highest-first selection

rule the active or violating vertex with the highest d-value is chosen. Due to Derigs & Meier (1989), the corresponding variant of the

Goldberg-algorithm is of complexity $O(n^3)$.

The different algorithms were tested using a series of graphs with different sizes and densities generated by NETGEN (cf. Klingman et al. 1974) and RMFGEN (cf. Goldfarb & Grigoriadis 1988). While the first generator is well known, we give a description of the main features of the RMFGEN-generator:

This program accepts values for four parameters a, b, c_1 and c_2 ($>c_1$), and generates networks with the structure depicted in Figure 2.5.

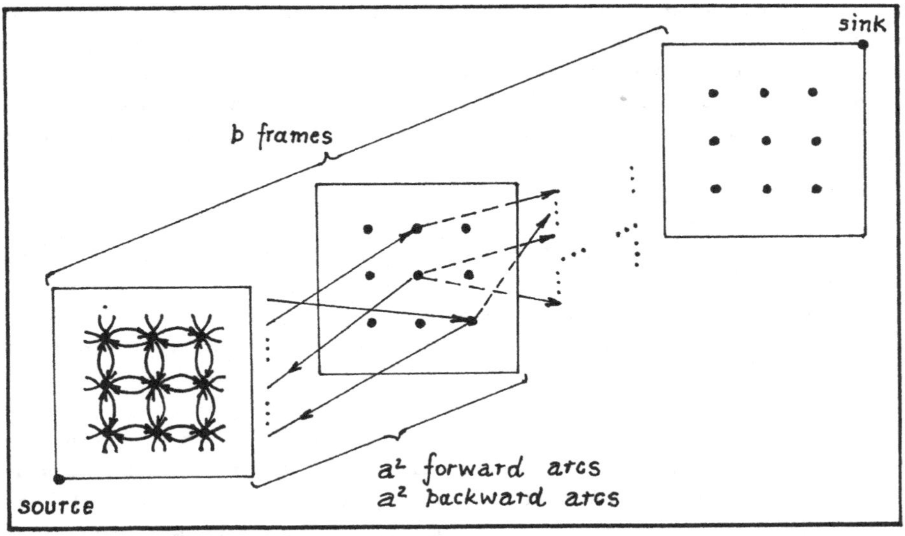

Figure 2.5. RMFGEN-graph structure.

The graph can be visualized as having b frames, each of which has a^2 vertices. Each frame is a connected subgraph with $4a \cdot b(a - 1)$ 'in-frame' arcs of capacity equal to $c_2 \cdot a^2$. There are a^2 arcs from the vertices of one frame to a pseudorandom permutation of the vertices of the two neighbored frames (one to one). The arcs have pseudorandom integer capacities in the range $[c_1, c_2]$. The generated graphs have $n = a^2 \cdot b$ vertices and $m = 6a^2 \cdot b - 4a \cdot b - 2a^2$ arcs.

The solution of RMFGEN-generated problems involves flow augmenting paths whose length is at least b. The minimum cut is located between two consecutive frames and it contains a^2 arcs. This is an advantage when compared with NETGEN-generated maximum flow problems often having minimum cuts with a small number of arcs located near the source or near the sink.

When comparing the different implementations of the Goldberg-algorithms it turned out that the optimal combinations for the RMFGEN-generated graphs and the NETGEN-generated graphs were not the same. In the first case (RMFGEN), the best implementation uses the exact start-labeling and the highest first selection rule in both phases. For the NETGEN-examples, the best implementation uses the naive startlabeling and LIFO-selection rule in phase I and the exact startlabeling and highest first selection in phase II. In both test series, gap-search was favorable in all cases and on every single example.

Secondly, the best implementations of the Goldberg-algorithm called GOLDRMF and GOLDNET as described above were tested against four implementations of the classical Dinic-algorithm and an implementation of Karzanov's algorithm.

The memory requirements of all these codes are given in Table 2.3. All codes were written in ANSI-FORTRAN and compiled using VMS-FORTRAN.

Table 2.3. Memory requirements of max-flow codes due to Derigs & Meier (1988).

Code	Author	Memory
DINIC-I	Imai (1983)	5n + 6m
DINIC-K	Koeppen(1980)	4n + 7m
DINIC-G	Goldfarb & Grigoriadis (1988)	6n + 5m
DINIC-T	the authors' implementation of the Dinic-Tarjan-algorithm	8n + 10m
KARZANOV-I	Imai (1983)	6n + 8n
GOLDRMF	Derigs & Meier (1989)	8n + 6m
GOLDNET	Derigs & Meier (1989)	8n + 6m

The main result of the computational study of Derigs & Meier (1989) is the superiority of the best Goldberg implementation over the other codes. To give a better impression, we report some of their average running times in CPU seconds on a VAX 8600 in Tables 2.4. and

2.5., respectively. Therein, the density den is defined to be den := 2m/n(n-1).

Table 2.4. Running times of different max-flow-codes on NETGEN-generated graphs.

n	den	DINIC-G	DINIC-I	DINIC-K	DINIC-T	KARZA-I	GOLDNET	GOLDRMF
300	3%	2.328	1.844	1.837	2.844	2.052	0.152	1.477
	5%	1.996	1.621	1.600	2.534	1.899	0.231	1.537
	10%	1.725	1.540	1.414	2.199	1.855	0.469	1.554
	20%	1.775	1.828	1.480	2.342	2.153	0.816	1.613
	40%	2.439	2.657	1.897	2.867	3.130	1.811	2.432
	60%	2.570	2.518	2.407	3.573	3.109	2.889	3.026

Table 2.5. Running times of different max-flow-codes on RMFGEN-generated graphs.

A	B	DINIC-G	DINIC-I	DINIC-K	DINIC-T	KARZA-I	GOLDNET	GOLDRMF
8	2	0.182	0.148	0.182	0.249	0.199	0.114	0.170
	4	0.574	0.457	0.504	0.690	0.566	0.233	0.585
	8	1.782	1.406	1.521	1.986	1.722	0.721	2.028
	16	5.190	4.156	4.119	5.783	5.236	1.799	5.314
	32	13.493	11.853	11.258	15.994	14.823	3.866	17.233
	64	46.032	40.357	38.573	57.598	48.666	8.841	57.920

2.6. Characterization of all Optimal Solutions

For many applications it is important to know more than only one optimal solution. We give a method to describe X_{max} or a subset of X_{max}, which is based on the knowledge of different minimum cuts and one maximum flow $x \in X_{max}$.

Lemma 2.8.
Any cut (X,X^*) is a minimum cut if and only if all flows $x \in X_{max}$ fulfill the property

$$(10) \quad x(i,j) = \begin{cases} cap(i,j) & \text{for } (i,j) \in \delta^+(X) \\ 0 & \text{for } (i,j) \in \delta^-(X). \end{cases}$$

Proof:
If (10) is valid for all $x \in X_{max}$ then

$$\begin{aligned}\#(x) &= \Sigma_{(i,j)\in\delta+(X)}\, x(i,j) - \Sigma_{(i,j)\in\delta-(X)}\, x(i,j)\\ &= \Sigma_{(i,j)\in\delta+(X)}\, cap(i,j)\\ &= Cap(X,X^*),\end{aligned}$$

i.e., $(X,X^*) \in C_{min}$. On the other hand , from the definition of $Cap(X,X^*)$ and the validity of the max-flow min-cut relation (Theorem 2.1.) we can confirm the validity of the second part of the proposition.

■

As a consequence of Lemma 2.8. we formulate

Lemma 2.9.
If $(X1,X1^*)$, $(X2,X2^*)$ are minimum cuts then also $(X1 + X2,X1^* \cap X2^*)$ and $(X1 \cap X2,X1^* + X2^*)$ are so either.

■

Now we assume that C_{min} or a subset C_{sub} are given. These sets can be obtained from an application of the results of Jarvis & Tufekci (1982). The use of Lemma 2.9. confirms the existence of a smallest and of a largest set X such that $(X,X^*) \in C_{min}$.

Let $V = V_1 + V_2 + ... + V_p$ be a partition of V such that with $X^q := V_1 + ... + V_q$ the following conditions are fulfilled:

(i) $X^1 \subset X^q$ for all cuts $(X^q,X^{q*}) \in C_{sub}$
(ii) $X^q \subset X^p$ for all cuts $(X^q,X^{q*}) \in C_{sub}$
(iii) each V_q, $2 \le q \le p-1$ has the representation
$V_q = X - Y$ with $(X,X^*),(Y,Y^*) \in C_{min}$.

The subgraph induced by V_q, $1 \le q \le p$ is denoted by $G_q := (V_q,A_q) = (V_q,A \cap (V_q \times V_q))$. For the graph shown in Figure 2.6. we assume a partition of V into the sets $V_1 = \{1,2,3,4\}$, $V_2 = \{5,8,9\}$, $V_3 = \{6,7,10,11\}$, and $V_4 = \{12,13,14\}$. Then four minimum cuts of the set C_{sub} are defined according to

$X^1 = \{1,2,3,4\}$,
$X^2 = \{1,2,3,4,5,8,9\}$,
$X^3 = \{1,2,3,4,6,7,10,11\}$,
$X^4 = \{1,2,3,4,5,6,7,8,9,10,11\}$.

The corresponding induced subgraphs are shown in Figure 2.7. With

$$A_0 := A - \sum_{k=1}^{p} A_k$$ and the feasible areas dX_q; $q = 1,\ldots,p$ defined

by

(11) $\sum_{(j,k)\in A} dx(j,k) - \sum_{(i,j)\in A} dx(i,j) = 0$ for all $j \in V_q$,
(12) $-x(i,j) \le dx(i,j) \le cap(i,j) - x(i,j)$ for all $(i,j) \in A_q$

we are able to describe the set X_{max}.

Theorem 2.4.
Let $x \in X_{max}$. Then all maximal flows $x^k \in X_{max}$ are of the form

$$x^k(i,j) = \begin{cases} x(i,j) + dx^k(i,j) \text{ with } dx^k \in dX_q & \text{for } (i,j) \in A_q \\ x(i,j) & \text{for } (i,j) \in A_0 . \end{cases}$$

Proof:
We consider the difference flow $dx^k = x^k - x$. From the flow property of x and x^k and $\#(x) = \#(x^k)$ it follows (11) for all vertices $j \in V$. (12) is an immediate consequence of the feasibility of x. From Lemma 2.8. we conclude that $x_k(i,j) = x(i,j)$ for all arcs $(i,j) \in A_0$.
■

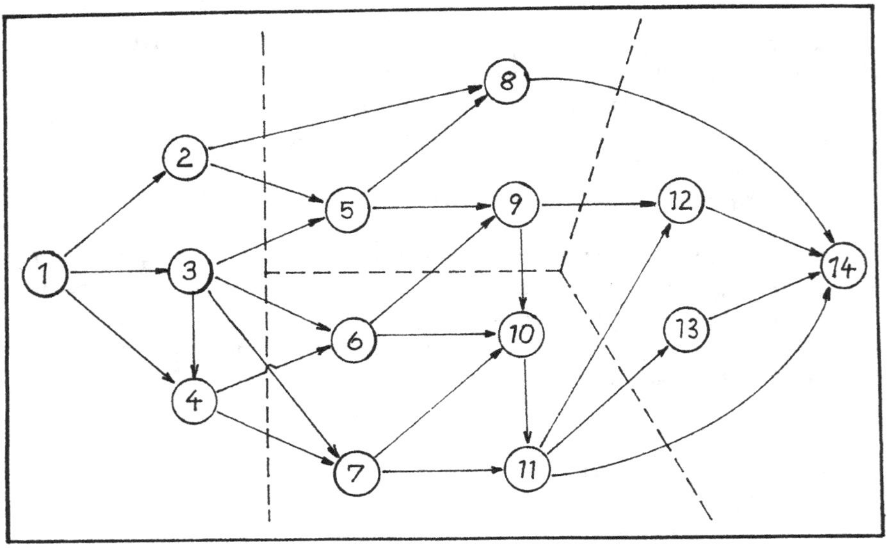

Figure 2.6. Graph G = (V,A) with indicated minimum cuts.

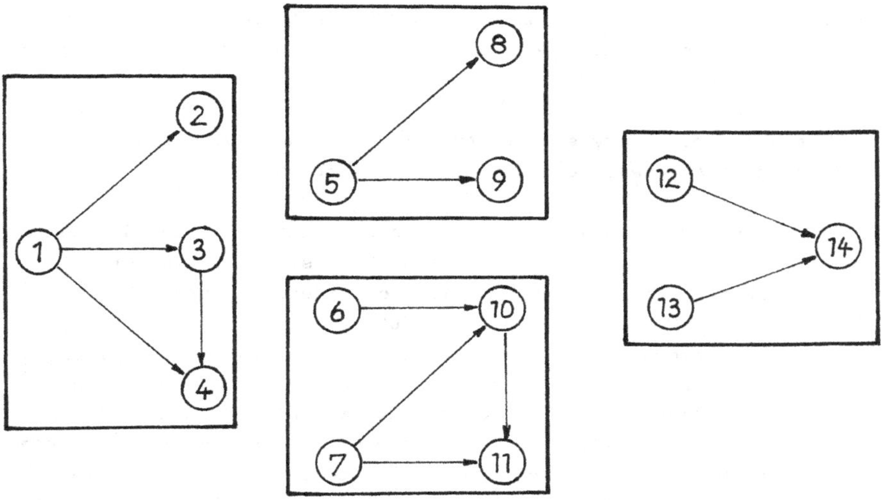

Figure 2.7. Decomposition of G into subgraphs G_q; q = 1,...,4.

By means of Theorem 2.4. we can describe the optimal solutions of MF using feasible solutions of corresponding subproblems. In the case of the above example we assume a maximal flow resulting in the following capacity intervals according to (12):

G_1	(1,2):	[0,2]	G_2	(5,8):	[-3,0]
	(1,3):	[-2,0]		(5,9):	[0,5]
	(1,4):	[0,4]			
	(3,4):	[0,1]			
G_3	(6,10):	[0,2]	G_4	(10,14):	[0,2]
	(7,10):	[-3,0]		(13,14):	[0,3].
	(7,11):	[0,3]			
	(10,11):	[-2,0]			

For subgraphs G^1 and G^3 there are two basic solutions in both cases. The first one is the zero flow. The remaining solutions result from a circulation of two respectively one unit along the cycle given by the vertices 1, 3, 4 and 7, 10, 11.

There are necessary some further remarks concerning the practical realization of the described decomposition method:

(i) The efficiency of the approach depends on the degree of the decomposition of G into subgraphs G_k. In the worst case there is an exponential dependence between $\#(C_{min})$ and n (Picard & Queyranne 1980).

(ii) To describe the feasible solutions of the subproblems we can restrict to basic solutions. If the subgraph is sufficiently small we can use an algorithm of Gallo & Sodini (1979) to enumerate all neighbored extreme flows (flows which correspond to extreme points in the polyhedron **X**). The adjacency relationship is based on the adjacency of the extreme points and the correspondence be-between extremal flows and spanning trees.

(iii) From a practical point of view it is sufficient to know a subset of C_{min}.

As a consequence of Theorem 2.4. we get a sufficient condition that a given maximal flow is the unique optimal solution.

Corollary
Assume that $x \in X_{max}$. If there is a decomposition of G into subgraphs G^q such that all G^q represent spanning trees, then x is the unique optimal solution.

Proof:
There is no possibility to get a nontrivial solution $dx \neq 0$ for any of the feasible areas dX_q ; $q = 1,\ldots,p$.

■

2.7. Maximal Flows between all Pairs of Vertices

In the context of the design and the analysis of communication and transportation networks, maximum flows between all pairs of vertices of an undirected graph G = (V,E) are often needed. We use the notation [i,j] for the undirected pair i,j contained in E and call E the set of edges. As before, each edge $[i,j]$ has a positive capacity $cap[i,j]$. For any fixed pair $k,l \in V$ a flow vector x_{kl} on E is defined. The maximum flow between two vertices k,l is abbreviated by $f(k,l)$. From symmetry it follows that $f(k,l) = f(l,k)$. The constraints

$$0 \le x_{kl}[i,j] \le cap[i,j] \text{ for all } [i,j] \in E$$

$$\Sigma_{(i,j)\in E}\, x_{kl}[i,j] - \Sigma_{(h,i)\in E}\, x_{kl}[h,i] = \begin{cases} \#(x_{kl}) & \text{for } i = k \\ -\#(x_{kl}) & \text{for } i = l \\ 0 & \text{otherwise} \end{cases}$$

are summarized in the polyhedron $X[k,l]$.

MTMF $\max\ \{\#(x_{kl}): x_{kl} \in X[k,l]\}$ for all pairs $k,l \in V$.

The notion of a minimum cut is transferred to the undirected case. The capacity of a cut (X,X^*) is given by

$$Cap[X,X^*] := \Sigma_{[i,j]\in[X,X^*]}\, cap[i,j].$$

A *minimum cut* is a cut of minimum capacity. $C[k,l]$ stands for the set of all cuts separating k and l. $C_{min}[k,l]$ denotes the subset of all minimum cuts in $C[k,l]$. Since each edge $[i,j] \in E$ can be replaced by two arcs $(i,j),(j,i)$ with capacity constraints due to (2), we obtain the validity of the max-flow min-cut theorem

$$(13) \quad \max\ \{\#(x_{kl}): x_{kl} \in X[k,l]\ \} = \min\ \{Cap[X,X^*]: (X,X^*) \in C[k,l]\}.$$

Two graphs $G = (V,E)$ and $G^1 = (V,E^1)$ with capacities cap respectively cap^1 are said to be *flow equivalent* if their solutions are the same. Gomory & Hu constructed a tree T and a weight function w defined on the edges of T such that for any two vertices $k,l \in V$, if e is the minimum weight edge on the path from k to l in T, then $f(k,l) = w(e)$. Such a tree is called an *equivalent flow tree*. In addition they proved that there exists an equivalent flow tree such that for every pair $k,l \in V$ with e as defined above, then the two components of $T - e$ form a minimum cut between k and l. Such a tree is called a (Gomory-Hu) *cut tree*. For an illustration of these concepts, consider the network shown in Figure 2.8.(a) and the corresponding equivalent flow tree T with weight function w of Figure 2.8.(b). As an example, the maximum flow $f(4,5)$ between 4 and 5 can be obtained from the minimum weight edge $e = [1,2]$, and it is $f(4,5) = w(1,2) = 11$.

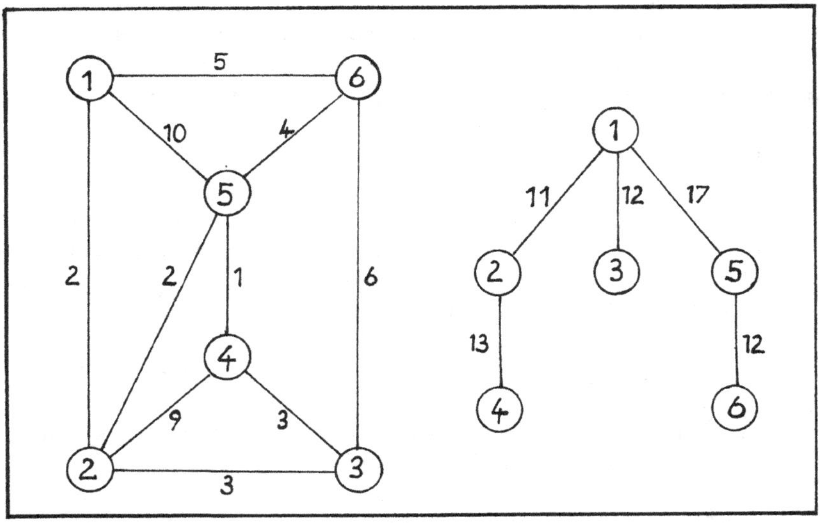

Figure 2.8. Graph G = (V,E) with capacities cap (left) and equivalent flow tree T with function w (right).

Let $(X,X^*),(Y,Y^*)$ be two cuts dividing $V = X + X^* = Y + Y^*$ into two disjoint subsets. The cuts (X,X^*) and (Y,Y^*) are said to *cross* each other iff each of the intersecting sets $X \cap Y$, $X \cap Y^*$, $X^* \cap Y$ and $X^* \cap Y^*$ contains at least one vertex.

For the solution of MTMF it was shown by Gomory & Hu (1961) that the $n(n-1)/2$ maximum flow values can be computed with only $n - 1$ maximum flow computations. The method consecutively contracts vertices of the underlying network. For a subset V_i of V, the contraction of V_i is the replacement of the vertices of V_i by a single vertex i, and for each $j \in V - V_i$, the replacement of the edges from j to V_i with a single edge between i and j. The capacity of this (new) edge between i and j is the sum of the capacities of the removed edges incident with j.

Theorem 2.5. (Gomory & Hu 1961)
A set of numbers $f(k,l)$; $k,l \in \{1,\ldots,n\}$ represents the maximum flow values of MTMF if and only if

(14) $f(k,l) \geq \min\{f(k,h),f(h,l)\}$ for all triples $h,k,l \in \{1,\ldots,n\}$. ∎

Theorem 2.6. (Gomory & Hu 1961)
Let $(X,X^*) \in C_{min}[i,j]$, $(Y,Y^*) \in C_{min}[k,l]$ be two crossing minimum cuts. Then there exist two minimum cuts, one separating i and j, the other separating k and l, such that these cuts do not cross each other.

■

Corollary.
To find a minimum cut separating two vertices i and j in X, all vertices in X^* can be contracted to a single vertex.

■

Most of the work in the Gomory-Hu method is involved in explicitly maintaining the non-crossing condition. In particular, the operations of vertex contraction and identification of which vertices to contract, are consequences of the need to maintain non-crossing cuts. Gusfield (1987) showed that vertex contraction is not needed, nor are non-crossing cuts explicitly needed. As a result, Gusfield presented two easy methods to determine an equivalent flow tree respectively a cut tree. An important property is the fact that the only interaction with graph G occurs with a routine producing a minimum cut. We investigate one of the methods in more detail.

Procedure GUSFIELD-EQ determines an equivalent flow tree T that is not necessarily a cut tree. At the beginning, T is a star with vertex 1 at the center and all remaining vertices 2 through n at the leaves. At each iteration, vertex s and all the vertices larger than s are leaves and have an unique neighbor in T. For s and t = p(s), the minimum cut is computed via a call of an oracle algorithm producing a minimum cut between two vertices in a capacitated network.

```
procedure GUSFIELD-EQ
begin
C p denotes the predecessor vector for the vertices in T
  for i := 1 to n do p(i) := 1
  for s := 2 to n do
  begin
    t := p(s)
    compute (X,X*) ∈ Cmin[s,t] with s ∈ X
    f(s,t) := Cap[X,X*]
    for i := s + 1 to n do
    begin
      if i ∈ X and p(i) = t then p(i) := s
C Assume F to be an n x n vector holding the flow values
      F(s,t) := f(s,t)
      F(t,s) := F(s,t)
      for i := 1 to s - 1 do
        if i ≠ t then
                begin
                  F(s,i) := min {f(s,t),f(t,i)}
                  F(i,s) := F(s,i)
                end
    end
  end
end
```

The algorithm is illustrated by the network shown in Figure 2.8. At the first iteration, the minimum cut between s = 2 and t = p(2) = 1 is calculated. The cut is given by X = {2,4}. The results of the subsequent iterations are summarized in Table 2.6. The final flow equivalent tree is shown in Figure 2.8.

Table 2.6. Iterations of GUSFIELD-EQ.

	s	t	X	X*	f(s,t)	p
1	2	1	{2,4}	{1,3,5,6}	11	(1,1,1,2,1,1)
2	3	1	{3}	{1,2,4,5,6}	12	(1,1,1,2,1,1)
3	4	2	{4}	{1,2,3,5,6}	13	(1,1,1,2,1,1)
4	5	1	{2,3,4,5,6}	{1}	17	(1,1,1,2,1,5)
5	6	5	{6}	{1,2,3,4,5}	12	(1,1,1,2,1,5)

In his second algorithm, Gusfield (1987) modified the original Gomory-Hu method to produce a cut tree. We have tested procedure GUSFIELD-EQ by a series of randomly generated examples varying in the number of vertices and the density of the graphs. The capacities have been generated randomly in the interval [1,100]. As an oracle for producing the minimum cut at each iteration, an implementation of Klinz (1988) of the Goldberg algorithm combined with gap search was used. The program was written in Turbo-Pascal. The computational results on a personal computer PC 1750 are summarized in Table 2.7. The given numbers are average values of five calculations performed in each class of test problems.

Table 2.7. cpu-times in seconds for testing GUSFIELD-EQ.

Problem	n	m	cpu
1	50	100	20.7
2	50	200	25.0
3	50	500	45.6
4	50	1000	85.5
5	100	200	81.0
6	100	500	108.1
7	100	1000	148.7
8	100	1500	250.3
9	100	2000	292.8
10	100	2500	388.8

§3 MINIMUM-COST FLOW PROBLEMS

3.1. Problem Statement and Fundamental Results

The minimum-cost flow problem defined on a directed graph $G = (V,A)$ is that of finding a feasible flow of minimum cost. In addition to the maximum flow problem, each arc $(i,j) \in A$ has associated an integer $c(i,j)$ referred to as cost per unit of flow. Let $b: V \longmapsto \mathbf{R}$ be the demand-supply vector, where $b(j) < 0$ if j is an origin vertex, $b(j) > 0$ if j is a destination vertex, and $b(j) = 0$ for a transshipment vertex j. Using the vertex-arc incidence matrix $I(G)$ of a graph, the minimum-cost flow problem may be written in the form

MCF: $\min \{c^T x: I(G)\cdot x = b,\ 0 \le x \le cap\}$.

Many network optimization problems are special cases of MCF. Among them there is the maximum flow problem, the shortest path problem, the maximum (weighted or cardinality) matching in bipartite graphs, and the transportation problem.

To formulate the dual problem MCF^d we associate dual variables $p(i)$ for all vertices $i \in V$ in correspondence to the flow conservation rule. Additionally, dual variables $y(i,j)$ are introduced in relation to the capacity constraints for all $(i,j) \in A$.

$$\mathbf{MCF^d}: \max \{b^T p - cap^T y : \quad p(j) - p(i) - y(i,j) \le c(i,j) \text{ for all } (i,j) \in A, \quad y(i,j) \ge 0 \text{ for all } (i,j) \in A\}.$$

The dual vector p in MCF^d is called *potential* or *price vector*. Given any price vector p we consider the corresponding *tension vector* t defined by $t(i,j) := p(j) - p(i)$ for all $(i,j) \in A$.

Theorem 3.1.
(x,p) is a primal and dual optimal solution pair if and only if
(1) $x(i,j) = 0$ for all $(i,j) \in A$ with $t(i,j) < c(i,j)$
(2) $0 \le x(i,j) \le cap(i,j)$ for all $(i,j) \in A$ with $t(i,j) = c(i,j)$
(3) $x(i,j) = cap(i,j)$ for all $(i,j) \in A$ with $t(i,j) > c(i,j)$
(4) $d(j) := b(j) + \Sigma_{(j,k)\in A}\, x(j,k) - \Sigma_{(i,j)\in A}\, x(i,j) = 0$ for all $j \in V$.

Proof:
The complementary slackness conditions of linear programming are:

$$y(i,j)[cap(i,j)-x(i,j)] = 0 \quad \text{for all } (i,j) \in A$$
$$x(i,j)[c(i,j) + p(i) - p(j) + y(i,j)] = 0 \quad \text{for all } (i,j) \in A.$$

Consequently, $t(i,j) < c(i,j)$ implies $x(i,j) = 0$, and $t(i,j) > c(i,j)$ implies $y(i,j) > 0$ and $x(i,j) = cap(i,j)$. The primal feasibility is expressed by (2) and (4). The dual feasibility is valid with $y(i,j)$ defined as $y(i,j) := \max \{0, t(i,j) - c(i,j)\}$.

■

An arc satisfying the complementary slackness conditions is said to be *in kilter*. In the classical out-of-kilter method independently developed by Fulkerson (1961) and Minty (1960), kilter numbers $K(i,j)$ are introduced to indicate the absolute value of the changes in $x(i,j)$ necessary to bring arc (i,j) into kilter. Specifically,

$$K(i,j) := \begin{cases} abs(x(i,j)) & \text{for } t(i,j) < c(i,j) \\ \max\{-x(i,j), x(i,j)-cap(i,j), 0\} & \text{for } t(i,j) = c(i,j) \\ abs(cap(i,j)-x(i,j)) & \text{for } t(i,j) > c(i,j) \end{cases}$$

The out-of-kilter algorithm iteratively reduces the kilter numbers until their sum is equal to zero. At each iteration, either a primal or a dual change in the flows respectively in the prices is made. A bound on the overall running time is $O(n^2K_0)$ where K_0 is the initial sum of all kilter numbers.

For any flow $x \in \mathbf{X}$ and the corresponding residual graph $R(x)$ as defined in chapter 2.1., costs

$$(5) \quad c(i,j) := \begin{cases} c(i,j) & \text{for } (i,j) \in A^+(x) \\ -c(j,i) & \text{for } (i,j) \in A^-(x) \end{cases}$$

and the residual capacity

$$(6) \quad rcap(i,j) := \begin{cases} cap(i,j) - x(i,j) & \text{for } (i,j) \in A^+(x) \\ x(j,i) & \text{for } (i,j) \in A^-(x) \end{cases}$$

are defined. A directed cycle L in $R(x)$ indicates that there may be some flow augmentation along L. The weight $c(L)$ in $R(x)$ is defined due to $c(L) := \Sigma_{(i,j)\in L}\, c(i,j)$.

Theorem 3.2. (Klein 1967)
A flow $x \in \mathbf{X}$ is an optimal solution of MCF if and only if there is no cycle L in R(x) with $c(L) < 0$.

Proof:
Firstly, assume a negative cycle L in R(x). Then we can push flow around the cycle such that for the resulting flow x' it holds $c^T x' < c^T x$ and this contradicts the optimality of x.
Secondly, assume that there is no cycle L in R(x) with $c(L) < 0$ and x is not optimal, i.e., there is a flow x': $\#(x') = \#(x)$ and $c^T x' < c^T x$. Then consider $\delta x := x' - x$. The difference flow can be represented as a linear combination of flow along fundamental cycles. Since $c^T \delta x < 0$, at least one of these cycles must have negative cost, and this contradicts our assumption.

■

The above theorem is the foundation of what is called the negative cycle algorithm. It maintains a primal feasible solution x and strives to attain dual feasibility. It does so by identifying consecutively negative cost cycles in the residual network R(x) and augmenting flows in these cycles. The algorithm terminates if R(x) does not contain a negative cost cycle.

```
procedure NEGATIVE CYCLE
begin
  computation of a feasible flow x ∈ X
  while R(x) contains a negative cycle do
  begin
    detection of a negative cycle L
    ε := min {rcap(i,j): (i,j) ∈ L}
    x := x + ε·char(L)
  end
end
```

The negative cycle algorithm has attained new attraction in the recent past. This is due to some improvements in the way to identify a negative cost cycle. As in §2, U is used for upper bound on integral arc capacities. In addition we assume that the integral cost coefficients are bounded in absolute value by C. Identifying a negative cost cycle with maximum improvement

$$\Sigma_{(i,j) \in L}\, c(i,j) \cdot (\min \{rcap(i,j) : (i,j) \in L\})$$

in the objective function implies a method that would obtain the optimum flow within $O(C \cdot U \cdot m \cdot \log m)$ iterations. Though finding the cycle that gives the biggest improvement is NP-hard, a modest variation of this approach yields a polynomial time algorithm. The main point is that one can cancel vertex-disjoint cycles simultaneously based upon solving an auxiliary assignment problem (for further details compare Barahona & Tardos 1988).

A further improvement was achieved by identifying negative cost cycles with minimum mean cost. The mean cost mc(L) of a cycle L is defined as the cost of the cycle divided by the number of arcs it contains:

$$mc(L) := \Sigma_{(i,j)\epsilon L}\, c(i,j)/\#(L).$$

A *minimum mean cost cycle* is a cycle whose mean cost is as small as possible. It can be determined in $O(C \cdot n^{1/2} \cdot m \cdot \log n)$ time due to Orlin & Ahuja (1988). Goldberg & Tarjan (1988) showed that if flow is always augmented along a minimum mean cycle, then the negative cycle algorithm is strongly polynomial.

Theorem 3.3. (Busacker & Gowen 1961)
Let x1 be a minimum-cost flow among all flows x: $\#(x) = K_1$. If $P \in P(x1)$ is a shortest path from 1 to n in R(x1) with weights as defined in (5) then the flow

$$x2 := x1 + \epsilon \cdot char(P) \quad \text{with}$$
$$\epsilon = \min \{rcap(i,j): (i,j) \in P\}$$

is a minimum-cost flow among all flows $x \in \mathbf{X}$ with $\#(x) = K_1 + \epsilon$.

■

The above theorem suggests an algorithm, in which consecutively shortest paths are determined. Zadeh (1973) described a pathological example on which each of the following algorithms performs an exponentially (in dependence of n) long sequence of iterations:

- primal simplex algorithm with Dantzig's pivot rule,
- the dual simplex algorithm,
- the negative cycle algorithm (which augments flow along a most negative cycle),
- the successive shortest path algorithm according to Theorem 3.3,
- the primal-dual, and
- the out-of kilter algorithm.

For a given $\epsilon \geq 0$ a flow $x \in \mathbf{X}$ is called *ϵ-optimal* if there is a price vector $p: V \longmapsto \mathbf{R}$ such that the usual complementary slackness conditions are violated by at most ϵ:

$$t(i,j) - \epsilon > c(i,j) \quad \text{implies } x(i,j) = cap(i,j)$$
$$t(i,j) + \epsilon < c(i,j) \quad \text{implies } x(i,j) = 0.$$

Theorem 3.4. (Bertsekas 1986)
Suppose that $c(i,j) \in \mathbf{Z}$ for all $(i,j) \in A$. Then for any ϵ such that $0 \le \epsilon < 1/n$, an ϵ-optimal flow is also 0-optimal.

Proof:
Assume the contrary, i.e., that a given ϵ-optimal flow x is not 0-optimal. From Theorem 3.2. it follows the existence of a cycle L in R(x) with negative costs $c(L) < 0$. By the definition of ϵ - optimality and using $n \cdot \epsilon < 1$ implies

$$c(L) \ge \Sigma_{(i,j)\in L+} (t(i,j)-\epsilon) + \Sigma_{(i,j)\in L-} (-t(i,j)-\epsilon)$$
$$\ge \Sigma_{(i,j)\in L} char_{ij}(L)\cdot t(i,j) - n\cdot\epsilon > 0 - 1.$$

Since the costs are integers, the cost of the cycle must be at least 0. ■

```
procedure SUCCESSIVE APPROXIMATION
begin
  compute a feasible flow x ∈ X
  for all i ∈ V do p(i) := 0
  ε  := max {c(i,j): (i,j) ∈ A}
  repeat
  begin
    ε := ε/2
    [x_ε,p_ε] := REFINE[x_2ε,p_2ε,ε]
  end
  until ε < 1/n
end
```

The algorithm SUCCESSIVE APPROXIMATION of Goldberg & Tarjan (1987) starts by finding an approximate solution with $\epsilon = \max \{abs(c(i,j)): (i,j) \in A\}$. Each iteration begins with a triple $(\epsilon, x_\epsilon, p_\epsilon)$ such that x_ϵ is an ϵ-optimal flow with potential p_ϵ. At the end, a new triple $(\epsilon/2, x_{\epsilon/2}, p_{\epsilon/2})$ with the corresponding properties is found. This is done by forcing all edges "in kilter" and a subsequent transformation of the resulting vector (which need not necessarily satisfy (4)) into a $\epsilon/2$-optimal flow. The corresponding subroutine is called REFINE due to the original authors. When the error parameter is small enough, the current solution is optimal, and the algorithm terminates as a consequence of Theorem 3.4.

3.2. History of Polynomial Algorithms

The broad spectrum of activities for solution algorithms of MCF can be divided into two classes. The first is motivated primarily by the worst-case complexity , the second by empirical tests concerning running time for randomly generated examples. In Table 3.1., the best known (polynomial-time) algorithms and their complexity are summarized. For the solution of the shortest path and the maximum flow problems -both occurring as subroutines- the complexity bounds $O(m + n \cdot \log n)$ of Fredman & Tarjan (1984) respectively $O(m \cdot n \cdot \log(n^2/m))$ of Goldberg & Tarjan (1987) using the very sophisticated data structure of dynamic trees are used.

Table 3.1. History of polynomial algorithms for MCF

Edmonds & Karp (1972)	$O((m + n \cdot \log n) m \cdot \log U)$
Röck (1980)	$O((m + n \cdot \log n) m \cdot \log U)$
Röck (1980)	$O((n \cdot m \log(n^2/m)) n \cdot \log C)$
Tardos (1985)	$O(m^4)$
Orlin (1984)	$O(m^2 \cdot (\log n) \cdot (m + n \cdot \log n))$
Fujishige (1986)	$O(m^2 \cdot (\log n) \cdot (m + n \cdot \log n))$
Bland & Jensen (1985)	$O(n \cdot (\log C) \cdot (n \cdot m \log(n^2/m)))$
Galil & Tardos (1986)	$O(n^2 \cdot (\log n) \cdot (m + n \log n))$
Goldberg & Tarjan (1987)	$O(n \cdot m \cdot (\log n) \cdot \log(n \cdot C))$
Gabow & Tarjan (1987)	$O(n \cdot m \cdot (\log n)(\log U)(\log n \cdot C))$
Ahuja, Goldberg, Orlin & Tarjan (1988)	$O(n \cdot m \cdot \log \log U \log n \cdot C)$
Orlin (1988)	$O(m \cdot (\log n) \cdot (m + n \cdot \log n))$

For problems satisfying the similarity assumption (see Gabow 1985) that

$C = O(n^{O(1)})$ and $U = O(n^{O(1)})$

the double scaling algorithm of Ahuja, Goldberg, Orlin & Tarjan (1988) is faster than all other algorithms for all network topologies except for very dense graphs. Among the strongly polynomial algorithms the one of Orlin (1988) is the currently fastest. This algorithm solves the minimum-cost flow problem as a sequence of $m \cdot \log n$ shortest path problems. The theoretically bounds of the algorithms mentioned in Table 3.1 are not yet confirmed by computational experience proving the efficiency also for empirical test examples. Among the really implemented algorithms for solving MCF, Grigoriadis (1986) implementation of the network simplex method and the relaxation code of Bertsekas & Tseng (1986) are the favourites. The network simplex method is described in Section 3.3. in more detail. Computational results comparing both algorithms are presented in Section 3.4.

3.3. The Network Simplex Method

The network simplex method is a specialization of the bounded variable simplex method for linear programming enhanced by the use of appropriate data structures and various pivot selection rules. This specialization is of computational importance, since it completely eliminates the need for carrying and updating the basis inverse. Some of the most important foundations for the network simplex method are summarized in the following

Theorem 3.5.
Let $G = (V,A)$ be a connected graph with $\#(V) = n$.

(i) The vertex-arc incidence matrix $I(G)$ of G has rank $n - 1$.

(ii) The set of columns of $I(G)$ indexed by a subset $A_T \subset A$ of A is a column-basis of $I(G)$ iff $T = (V,A_T)$ is a spanning tree of G.

(iii) Let $I(T)$ be the node-arc incidence matrix of a spanning tree T. If $I(T)$ is made non-singular deleting one row, $I(T)$ can be permuted to a triangular matrix $B(T)$.

■

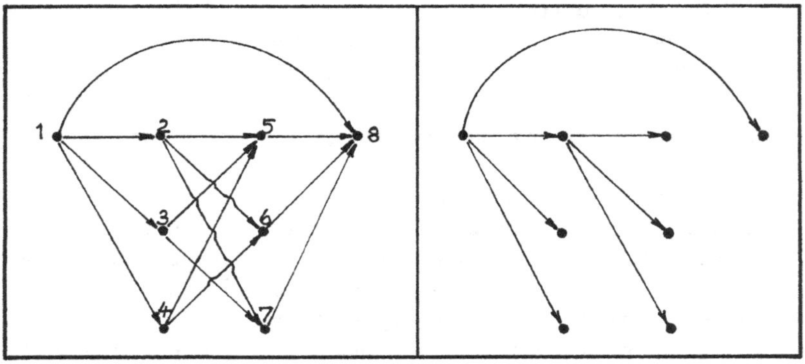

Figure 3.1. Graph $G = (V,A)$ with spanning tree $T = (V,A_T)$.

The propositions of the above theorem are illustrated by the graph shown in Figure 3.1. The related incidence matrix I(G) is:

(1,2)	(1,3)	(1,4)	(1,8)	(2,5)	(2,6)	(2,7)	(3,5)	(3,7)	(4,5)	(4,6)	(5,8)	(6,8)	(7,8)
-1	-1	-1	-1										
1				-1	-1	-1							
	1						-1	-1					
		1							-1	-1			
				1			1		1		-1		
					1					1		-1	
						1		1					-1
			1								1	1	1

Taking the spanning tree T of Figure 3.1. and deleting the last row we obtain I(T):

(1,8)	(1,2)	(1,3)	(1,4)	(2,5)	(2,6)	(2,7)
-1	-1	-1	-1			
	1			-1	-1	-1
		1				
			1			
				1		
					1	
						1

Let $x \in \mathbf{X}$ be a flow and $T = (V, A_T)$ a spanning tree. x is called a *basic solution* if for all $(i,j) \in A - A_T$ it holds either $x(i,j) = 0$ or $x(i,j) = cap(i,j)$. A basic solution is *feasible* if $0 \le x(i,j) \le cap(i,j)$ for all arcs $(i,j) \in A_T$. Finally, a basic solution is *degenerated*, if there is at least one $(i,j) \in A_T$ with either $x(i,j) = 0$ or $x(i,j) = cap(i,j)$.

The simplex algorithm maintains basic feasible solutions at each stage. The fundamental steps of the network simplex method without refinements as data structures for handling the trees or different pivot selection rules are presented in procedure SIMPLEX. Therein, the existence of a feasible basic solution $x \in \mathbf{X}$ related to a spanning tree $T = (V, A_T)$ and sets

$A_C := A - A_T$,

$A_{C-} := \{(i,j) \in A_C: x(i,j) = 0\}$, and

$A_{C+} := \{(i,j) \in A_C: x(i,j) = cap(i,j)\}$

is assumed.

```
procedure SIMPLEX
begin
  repeat
  COMPUTE PRICES
C Computation of the dual price vector
  for all (i,j) ∈ A_C do rc(i,j) := c(i,j) + p(i) - p(j)
  A_C-* := {(i,j) ∈ A_C-: rc(i,j) < 0}
  A_C+* := {(i,j) ∈ A_C+: rc(i,j) > 0}
  A_C* := A_C+* + A_C-*
  if A_C-* ≠ φ then
  begin
    choose (k,l) ∈ A_C-*
C Arc a = (k,l) is violating the optimality conditions
    CYCLE μ(T,a)
C Computation of the unique cycle μ(T,a) defined in T + {(k,l)}
    ε1 := min {x(i,j): (i,j) ∈ μ⁻(T,a)}
    ε2 := min {cap(i,j) - x(i,j): (i,j) ∈ μ⁺(T,a)}
    ε  := min {ε1,ε2}
C Assume that ε is determined by (s,t) ∈ μ(T,a)
    A_T := A_T + {(k,l)} - {(s,t)}
    x := x + ε·char(μ(T,a))
  end
  if A_C-* = φ and A_C+* ≠ φ then
  begin
    choose (k,l) ∈ A_C+*
    CYCLE μ(T,a)
    ε1 := min {cap(i,j) - x(i,j): (i,j) ∈ μ⁻(T,a)}
    ε2 := min {x(i,j): (i,j) ∈ μ⁺(T,a)}
    ε := min {ε1,ε2}
C Assume that ε is determined by (s,t) ∈ μ(T,a)
    A_T := A_T + {(k,l)} - {(s,t)}
    x := x + ε·char(μ(T,a))
  end
end
```

Subsequently, we describe the various steps performed by the network simplex method in greater detail. This includes a description of the two subroutines COMPUTE PRICES and CYCLE.

Comparable with linear programming, it needs some extra effort to compute a first basic solution. This can be done by introducing dummy arcs of very high costs.

One of the main reasons of the efficiency of the network simplex method is the way to represent the basic solutions . There are different possibilities to do this. We describe one such tree representation in more detail and consider the tree as "hanging" from a special vertex, the root. Then we use three functions defined on V:

- The predecessor function pred(i) gives for each $i \in V$ the first vertex in the unique path connecting i and the root:

$$\text{pred}(i) := \begin{cases} j & \text{if } (j,i) \in A_T \\ 0 & \text{if } i \text{ is the root of the tree.} \end{cases}$$

- The thread function thread(i) allows the visiting of the order; by convention the thread of the last vertex in the ordering is the root of the tree.
- The depth(i) function provides the length (number of arcs) of the path connecting i and the root.

The three functions are illustrated in Figure 3.2; the dotted lines indicate the sequence of vertices given by the thread function.

To select a candidate variable to enter the basis we need the price or potential vector. Since the rank of I(G) is n-1, the value of one potential can be set arbitrarily. We assume that p(1) = 0. The remaining potentials can be computed by using the condition

$$rc(i,j) = c(i,j) + p(i) - p(j) = 0 \quad \text{for all } (i,j) \in A_T$$

The thread indices allow us to compute all prices in O(n) time :

```
procedure COMPUTE PRICES
begin
  p(1) := 0
  j := thread(1)
  while j ≠ 1 do
  begin
    i := pred(j)
    if (i,j) ∈ A  then p(j) := p(i) - c(i,j)
    if (j,i) ∈ A  then p(j) := p(i) + c(i,j)
    j := thread(j)
  end
end
```

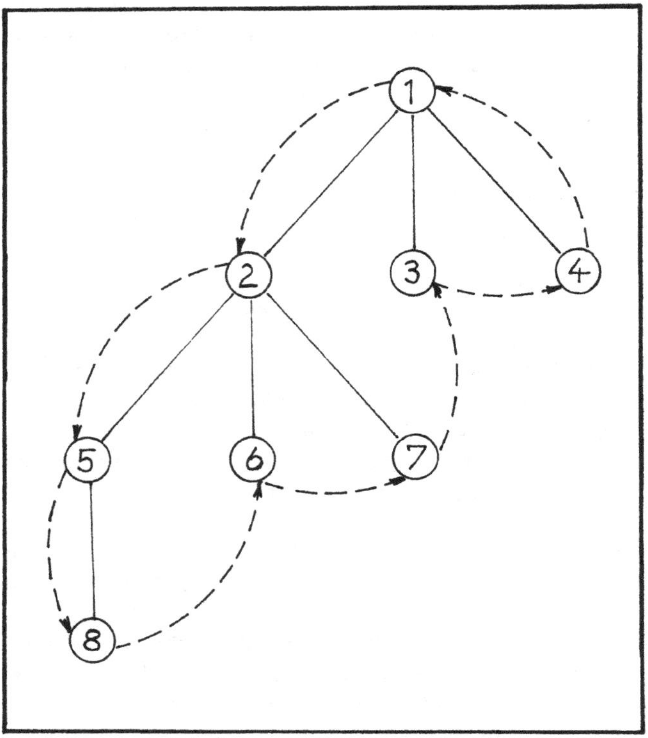

i	1	2	3	4	5	6	7	8
pred(i)	0	1	1	1	2	2	2	3
thread(i)	2	5	4	1	8	7	3	6
depth(i)	0	1	1	1	2	2	2	3

Figure 3.2. Example of a rooted tree and its tree indices.

Based on the dual vector p the reduced costs can be computed. Any nonbasic arc (i,j) at its lower bound with a negative reduced cost rc(i,j), or at its upper bound with a positive reduced cost, is eligible to enter the basis. From computational studies (compare Mulvey 1978) it is known that the method for selecting the entering arc has a major influence on the performance of the simplex algorithm. Different strategies for this choice were tested:

- "most negative", i.e., selection of the arc that violates the optimality condition the most,
- "first negative", i.e., selection of the first arc that violates the optimality condition when the arc list is examined cyclically at each iteration, and
- "candidate list approach" being an effective compromise between the two previous strategies and mostly used for implementations. In the candidate list, a set of arcs violating the optimality condition is maintained. When scanning the arcs the list is updated by removing those arcs that no longer violate the optimality condition. If the list becomes empty or a given limit of iterations for this list is reached, the list is rebuild.

Assume an arc $a = (k,l) \in A_C$ violating the optimality conditions is found. The set $A_T + \{(k,l)\}$ uniquely determines an elementary cycle $\mu(T,a)$ containing only arcs from A_T with the only exception of (k,l). The orientation of $\mu(T,a)$ is defined by (k,l): it is the same as that of (k,l) if $x(k,l) = 0$ and opposite to the orientation of (k,l) if $x(k,l) = cap(k,l)$. The following procedure CYCLE determines $\mu(T,a)$. Using the predecessor function the two paths P(k) and P(l) connecting the vertices k respectively l with the root 1 are identified. Then $\mu(T,a)$ is characterized by the set of arcs $\{(k,l)\} + P(k) + P(l) - (P(k) \cap P(l))$. Consequently, what remains to do is to find the first common ancestor of both k and l.

For the basis exchange step we ask for the bottleneck arc when pushing flow along the cycle $\mu(T,a)$. Suppose that the maximum flow update is determined by the arc $(s,t) \in A_T$. There are two possibilities: (s,t) equal or unequal to (k,l). In the first case the flow value on (k,l) moves from its upper to its lower bound, or vice versa. In the second case, adding (k,l) to the basis and deleting (s,t) from it again results in a basis which is a spanning tree.

How to update the potential vector? The deletion of (k,l) partitions the set V into two subtrees, one, T_1, containing the root, and the other, T_2, not containing the root. The conditions $rc(i,j) = 0$ for all $(i,j) \in A_T$ imply that the potentials of all vertices in T_1 remain unchanged. In the case of T_2, the change is by a constant amount $-rc(k,l)$ respectively $rc(k,l)$ depending on $k \in T_1$ (and $l \in T_2$) or $k \in T_2$ (and $l \in T_1$).

```
procedure CYCLE μ(T,a)
begin
  ε := cap(k,l)
  if x(k,l) = cap(k,l) then i := k
                            j := l
  if x(k,l) = 0        then i := l
                            j := k
  repeat
  begin
    if depth(i) > depth(j) then
    begin
      i1 := i
      i := pred(i)
      if (i,i1) ε A_T then  ε := min {ε,cap(i,i1) - x(i,i1)}
                     else  ε := min {ε,x(i1,i)}
    end
    if depth(j) > depth(i) then
    begin
      j1 := j
      j := pred(j)
      if (j1,j) ε A_T then  ε := min {ε,cap(j1,j) - x(j1,j)}
                     else  ε := min {ε,x(j,j1)}
    end
    if depth(j) = depth(i) then
    begin
      i1 := i
      i := pred(i)
      if (i,i1) ε A_T then  ε := min {ε,cap(i,i1)-x(i,i1)}
                     else  ε := min {ε,x(i1,i)}
      j1 := j
      j := pred(j)
      if (j1,j) ε A_T then  ε := min {ε,cap(j1,j)-x(j1,j)}
                     else  ε := min {ε,x(j,j1)}
    end
  end
  until i = j
end
```

We illustrate the network simplex method by the graph shown in Figure 3.1. with the following cost and capacity functions:

(i,j)	c(i,j)	cap(i,j)
(1,2)	0	1
(1,3)	0	3
(1,4)	0	5
(1,8)	100	100
(2,5)	0	100
(2,6)	1	100
(2,7)	3	100
(3,5)	1	100
(3,7)	3	100
(4,5)	3	100
(4,6)	3	100
(5.8)	0	2
(6,8)	0	2

SIMPLEX performs ten iterations which are reported in Figure 3.3. The demand-supply vector is b = (-8,0,0,0,0,0,0,8). In the first basic solution, all the eight flow units are transferred along the arc (1,8). The value of the (primal) objective function is decreased consecutively by pushing flow along flow augmenting paths which is cheaper than c(1,8).

The entries in the reduced price vector are in correspondence to the sequence of arcs given by A_{C+} and A_{C-}. With $\epsilon = 0$, the exchange between the second and the third basic solution is a degenerate one. For the final solution, the optimality conditions as stated in Theorem 3.1. are satisfied: The reduced costs of the arc (1,8) with flow x(1,8) is non-negative, while for all other non-basic arcs the corresponding flows are on the upper capacity bound and the reduced costs are non-positive. Additionally, the example originally introduced by Zadeh (1973) proves that the sequence of iterations caused by Dantzig's 'most negative' pivot rule may be of length exponentially depending on the number n of vertices.

1

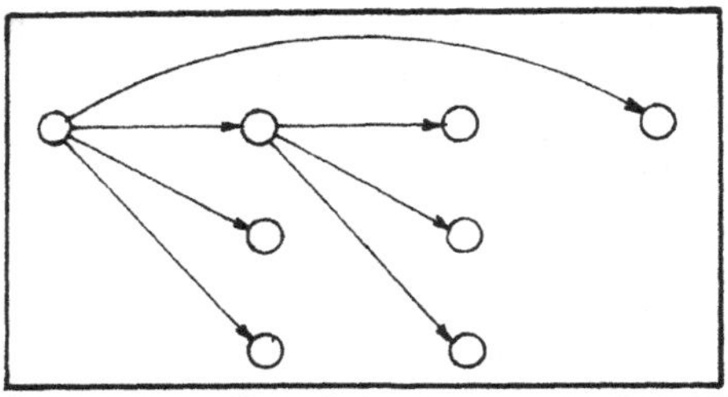

$p = (0,0,0,0,0,1,3,100)$
$A_{C+} = \emptyset$
$A_{C-} = \{(3,5),(3,7),(4,5),(4,6),(5,8),(6,8),(7,8)\}$
$rc = (1,0,3,2,-100,-99,-97)$
$(k,l) = (5,8), \ (s,t) = (1,2)$
$\epsilon = 1$

2

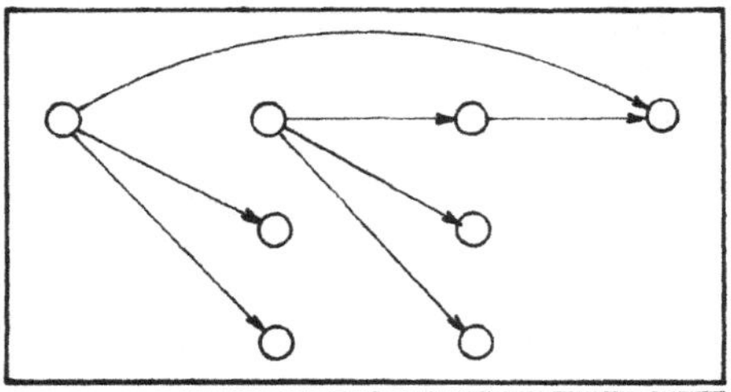

$p = (0,100,0,0,100,101,103,100)$
$A_{C+} = \{(1,2)\}$
$A_{C-} = \{(3,5),(3,7),(4,5),(4,6),(6,8),(7,8)\}$
$rc = (-100,-99,-100,-97,-98,1,3)$
$(k,l) = (3,7), \ (s,t) = (2,7)$
$\epsilon = 0$

3

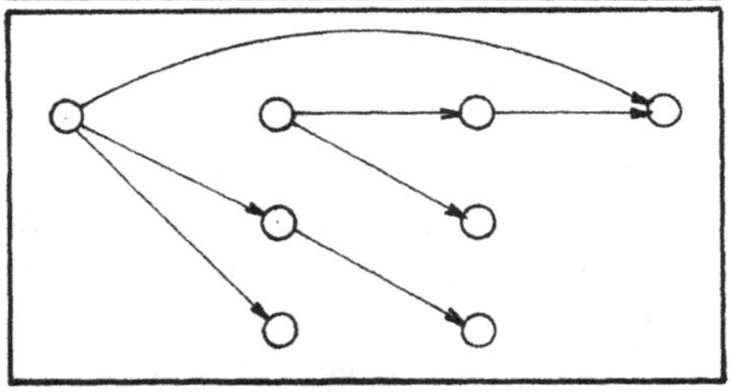

$p = (0,100,0,0,100,101,3,100)$
$A_{C+} = \{(1,2)\}$
$A_{C-} = \{(2,7),(3,5),(4,5),(4,6),(6,8),(7,8)\}$
$rc = (-100,100,-99,-97,-98,1,-97)$
$(k,l) = (3,5), \ (s,t) = (5,8)$
$\epsilon = 1$

4

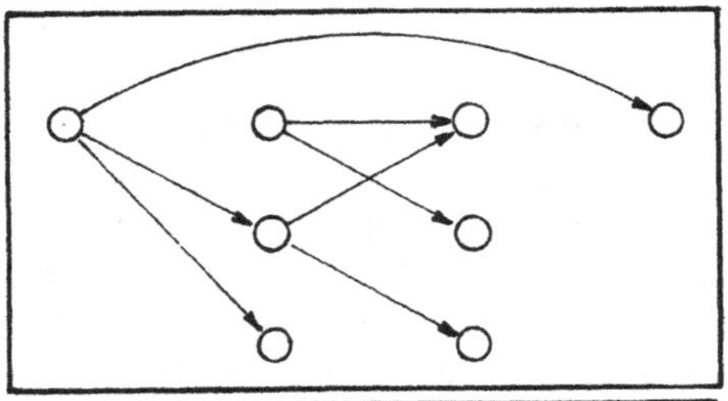

$p = (0,1,0,0,1,2,3,100)$
$A_{C+} = \{(1,2),(5,8)\}$
$A_{C-} = \{(2,7),(4,5),(4,6),(6,8),(7,8)\}$
$rc = (-1,-99,1,2,1,-98,-97)$
$(k,l) = (6,8), \ (s,t) = (2,5)$
$\epsilon = 1$

5

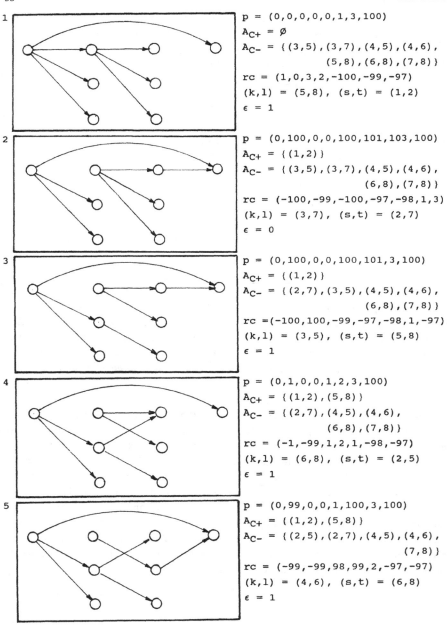

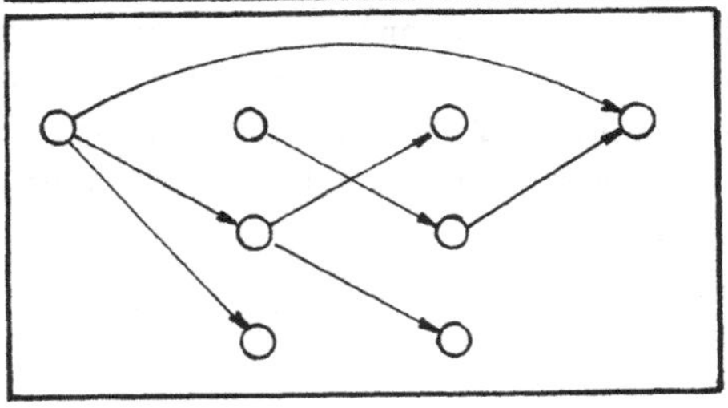

$p = (0,99,0,0,1,100,3,100)$
$A_{C+} = \{(1,2),(5,8)\}$
$A_{C-} = \{(2,5),(2,7),(4,5),(4,6),(7,8)\}$
$rc = (-99,-99,98,99,2,-97,-97)$
$(k,l) = (4,6), \ (s,t) = (6,8)$
$\epsilon = 1$

6

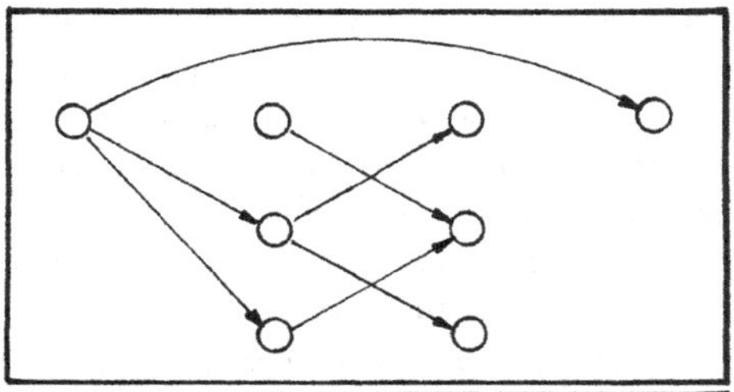

p = (0,2,0,0,1,3,3,100)

A_{C+} = {(1,2),(5,8),(6,8)}

A_{C-} = {(2,5),(2,7),(4,5),
(7,8)}

rc = (-2,-99,-97,1,2,2,-97)

(k,l) = (7,8), (s,t) = (1,3)

ϵ = 1

7

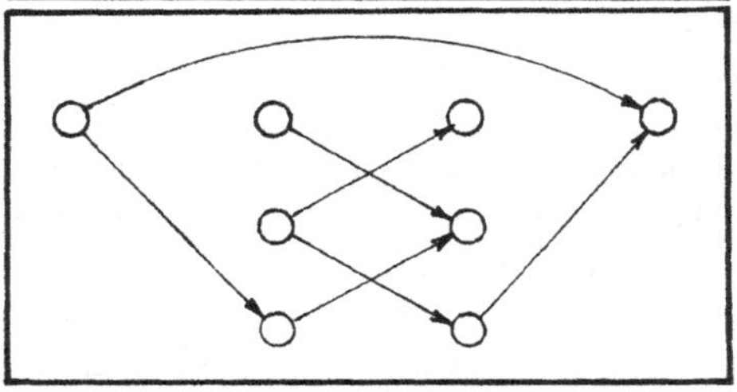

p = (0,2,97,0,98,3,100,100)

A_{C+} = {(1,2),(1,3),(5,8),(6,8)}

A_{C-} = {(2,5),(2,7),(4,5)}

rc = (-2,-97,-2,-97,-96,-95,-95)

(k,l) = (2,5), (s,t) = (2,6)

ϵ = 1

8

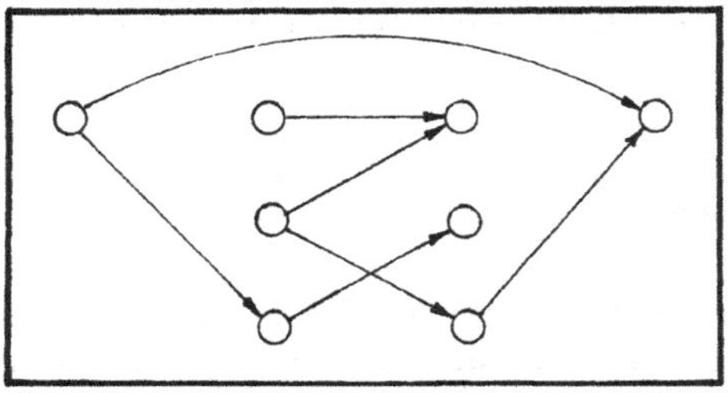

p = (0,98,97,0,98,3,100,100)

A_{C+} = {(1,2),(1,3),(5,8),(6,8)}

A_{C-} = {(2,6),(2,7),(4,5)}

rc = (-98,-97,-2,-97,96,1,-95)

(k,l) = (4,5), (s,t) = (3,5)

ϵ = 1

9

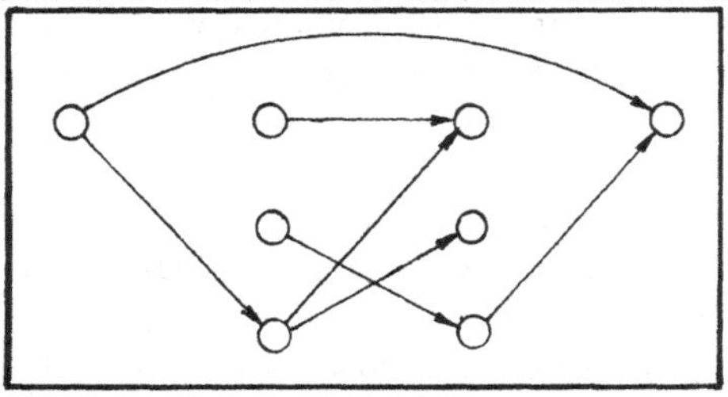

p = (0,3,97,0,3,3,100,100)

A_{C+} = {(1,2),(1,3),(5,8),(6,8)}

A_{C-} = {(2,6),(2,7),(3,5)}

rc = (-3,-97,-97,-97,1,-94,95)

(k,l) = (2,7), (s,t) = (1,8)

ϵ = 1

10

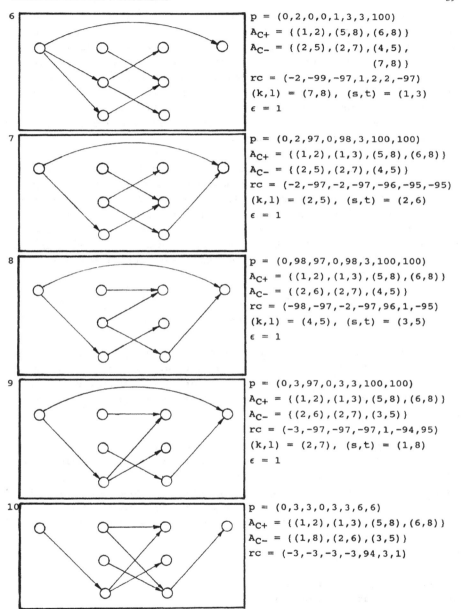

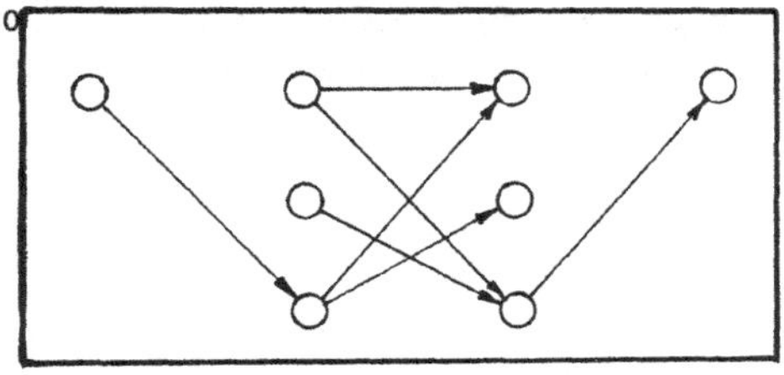

p = (0,3,3,0,3,3,6,6)

A_{C+} = {(1,2),(1,3),(5,8),(6,8)}

A_{C-} = {(1,8),(2,6),(3,5)}

rc = (-3,-3,-3,-3,94,3,1)

Figure 3.3. Subsequent iterations performed by SIMPLEX for the test example described above.

Finally, we investigate the question of termination. It is easy to show that the algorithm terminates in a finite number of steps if each pivot operation is nondegenerate, i.e., the ϵ value is greater than zero. One difficulty arising in simplex codes is the possibility of cycling caused by degenerate solutions. Due to examples of Bradley, Brown & Graves (1977), there have been more than 90% of the several 10^4 basic solutions degenerated. Cunningham (1976) introduced strongly feasible bases to avoid cycling, i.e., the repeated encountering of the same sequence of feasible bases.

Let T be a spanning tree and x its associated basic solution. We assume a fixed vertex $r \in V$ and assume the tree hanging from the root r. x is said to be *strongly feasible* if we can send a positive amount of flow from each vertex to r along arcs in the tree without violating any of the flow bounds. Since there is exactly one path in T to connect a vertex j and r, the definition implies that no upward pointing arc can be at its upper bound and no downward pointing arc can be at its lower bound.

Now the above procedure SIMPLEX is modified by a specific choice of the arc (s,t) entering T. Suppose that x(s,t) = 0.

> Let $v \in V$ be the first common ancestor in the paths in T from k and l to r. Choose (s,t) to be the last arc fixing ϵ when traversing μ(T,a) in the direction of (k,l) beginning at v (if x(s,t) = cap(s,t) then the cycle μ(T,a) is oriented opposite to (k,l)).

We denote by MSA (modified simplex algorithm) the fusion of SIMPLX with the above refinement and starting with a strongly feasible basis. Ahuja, Magnanti & Orlin (1988) showed that maintaining a strongly feasible basis is equivalent to applying the ordinary (network) simplex algorithm to the perturbed problem. Furthermore, they gave a bound for the number of pivot steps when using Dantzig's pivot rule, i.e., pivoting in the arc with maximum violation of the optimality conditions. Therein $W := m \cdot C \cdot U$ is used.

Theorem 3.6. (Ahuja, Magnanti & Orlin 1988)
MSA with the use of Dantzig's pivot rule performs $O(n \cdot m \cdot U \cdot \log W)$ pivots.

■

For the special cases of the shortest path and the assignment problem it is U = 1 and U = n, respectively. In fact, polynomial time simplex algorithms for the shortest path problem have been discovered by Dial et al. (1979), Zadeh (1979), Orlin (1985), Akgül (1985a), and Ahuja & Orlin (1988). In the case of the assignment problem, such algorithms were developed by Hung (1983), Orlin (1985), Akgül (1985b) and Ahuja & Orlin (1988). Developing a polynomial time primal simplex algorithm for the minimum-cost flow problem is still open. Orlin (1984) proved $O(n^3 \cdot \log n)$ pivots for the uncapacitated minimum-cost flow problem using a dual simplex algorithm.

3.4. Computational Results

A number of empirical studies have extensively tested different minimum-cost flow algorithms for a wide variety of network structures, data distributions and problem sizes. To have a common base for comparison, the problems are taken from the NETGEN generator developed by Klingman, Napier & Stutz (1974). This system is capable of generating assignment, capacitated or uncapacitated transportation and minimum-cost flow problems. Although it seems to be rather difficult to make a statement about the general performance of different algorithms, at the moment the relaxation algorithm of Bertsekas & Tseng (1988) and the primal simplex algorithm of Grigoriadis (1986) are appreciated to be the two fastest algorithms for solving the minimum-cost flow problem in practice. Both computer codes are available in the public domain.

In Table 3.2. we report some computational results of a more comprehensive study of Bertsekas & Tseng (1988). They compared their relaxation codes, RELAX-II and RELAXT-II, with RNET (a primal simplex code due to Grigoriadis & Hsu 1979) and KILTER. The last one is a primal-dual code due to Aashtiani and Magnanti (1976) and uses a sophisticated labeling scheme which is the fastest of nine versions tested by the authors. All methods were tested under identical conditions: same computer (VAX 11/750), same language (Fortran IV), compiler (standard Fortran of the VMS system version 3.7 in the Optimize mode), timing routine, and system conditions (empty system at night). All cpu-times do not include problem input and output. The results show the superiority of RELAX-II and RELAXT-II over the other codes for assignment and transportation problems. This is also valid for lightly capacitated and uncapacitated problems, where the margin of superiority increases for the large problems 37-40.

Table 3.2. Computational results due to Bertsekas & Tseng (1988) for NETGEN generated benchmark problems. All times are in cpu-seconds.

Problem Type #	n	m	RELAXT-II (VMS 3.7/ VMS 4.1)	RELAXT-II (VMS 3.7/ VMS 4.1)	KILTER VMS 3.7	RNET VMS 3.7
Transportation						
1	200	1300	2.07/1.75	1.47/1.22	8.81	3.15
3	200	2000	1.92/1.61	1.80/1.50	9.22	4.42
5	200	2900	2.97/2.43	2.53/2.05	16.48	7.18
6	300	3150	4.37/3.66	3.57/3.00	25.08	9.43
7	300	4500	5.46/4.53	3.83/3.17	35.55	12.60
8	300	5155	5.39/4.46	4.30/3.57	46.30	15.31
10	300	6300	4.12/3.50	3.78/3.07	47.80	16.44
Assignment						
11	400	1500	1.23/1.03	1.35/1.08	8.09	4.92
13	400	3000	1.68/1.42	1.87/1.54	8.99	8.92
15	400	4500	2.79/2.34	3.04/2.46	14.53	10.20
Uncapacitated and highly capacitated problems						
16	400	1306	2.79/2.40	2.60/2.57	13.57	2.76
17	400	2443	2.67/2.29	2.80/2.42	16.89	3.42
Uncapacitated and lightly capacitated problems						
28	1000	2900	4.90/4.10	5.67/5.02	29.77	8.60
29	1000	3400	5.57/4.76	5.13/4.43	32.36	12.01
30	1000	4400	7.31/6.47	7.18/6,26	42.21	11.12
31	1000	4800	5.76/4.95	7.14/6.30	39.11	10.45
32	1500	4342	8.20/7.07	8.25/7.29	69.28	18.04
33	1500	4385	10.39/8.96	8.94/7.43	63.59	17.29
34	1500	5107	9.49/8.11	8.88/7.81	72.51	20.50
35	1500	5730	10.95/9.74	10.52/9.26	67.49	17.81
Large uncapacitated problems						
37	5000	23000	87.05/73.64	74.67/66.66	681.94	281.87
38	3000	35000	68.25/57.84	55.84/47.33	607.89	274.46
39	5000	15000	89.83/75.17	66.23/58.74	558.60	151.00
40	3000	23000	50.42/42.73	35.91/30.56	369.40	174.74

§4 GENERALIZED NETWORKS

4.1. Maximum Flows in Generalized Networks

Generalized networks represent the most general class of network flow problems, which include capacitated and uncapacitated generalized transportation problems, the generalized assignment problem, as well as the corresponding network models without amplification. There is a great variety of practical problems which may be formulated in terms of generalized networks. Glover et al. (1978) mentioned applications in manufacturing, production, fuel to energy conversions, blending, crew scheduling, manpower to job requirements, and currency exchanges. The network-related formulation of all these problems allows a pictorial representation of the model, which is of advantage for the decision-maker. The special graph-theoretical structure of this class of linear programming problems has been employed to develop efficient solution methods. Using special data structures it is also possible to solve large-scale problems by these methods. The most effective way to represent generalized networks is the use of directed graphs. In addition to the notation introduced for pure networks, a positive value g(i,j) is associated with each arc $(i,j) \in A$. The main distinction between arcs in pure networks and arcs in generalized ones is that in the latter case the flow value is changed in passing the arc. In more detail, with x(i,j) units leaving vertex i, g(i,j)x(i,j) units reaching vertex j.

A vector $x \in \mathbf{R}^m$ is called a *generalized flow* if

(1) $\Sigma_{(i,j)\in A}\, g(i,j)\cdot x(i,j) - \Sigma_{(j,k)\in A}\, x(j,k) = 0$ for all $j \in V - \{1,n\}$

A generalized flow x is called *feasible* if

(2) $0 \le x(i,j) \le cap(i,j)$ for all $(i,j) \in A$,

i.e., the capacity constraints are related to the amount of flow entering the arc. The restrictions (1) and (2) define a polyhedron denoted $\mathbf{X}_g$.

Let $x_1 := \Sigma_{(1,k)\in A}\, x(1,k)$ and $x_n := \Sigma_{(i,n)\in A}\, g(i,n)\cdot x(i,n)$ be the amount of flow leaving the source respectively reaching the sink. A feasible generalized flow z is called a *max-min flow* if with respect to all $x \in \mathbf{X}_g$ it holds that

$$x_1 \leq z_1 \quad \text{implies} \quad x_n \leq z_n \quad \text{and}$$
$$x_n \geq z_n \quad \text{implies} \quad x_1 \geq z_1 \ .$$

A flow $z \in \mathbf{X}_g$ is called *n-maximum* if $z_n \geq x_n$ for all $x \in \mathbf{X}_g$. A flow z is called *n-optimal* if z is both a max-min flow and n-maximum. We illustrate the definitions by an example shown in Figure 4.1. Note that the arc flows are no longer necessarily integer. Additionally, the flow x_1 into the network is no longer equal to the flow x_n out of the network.

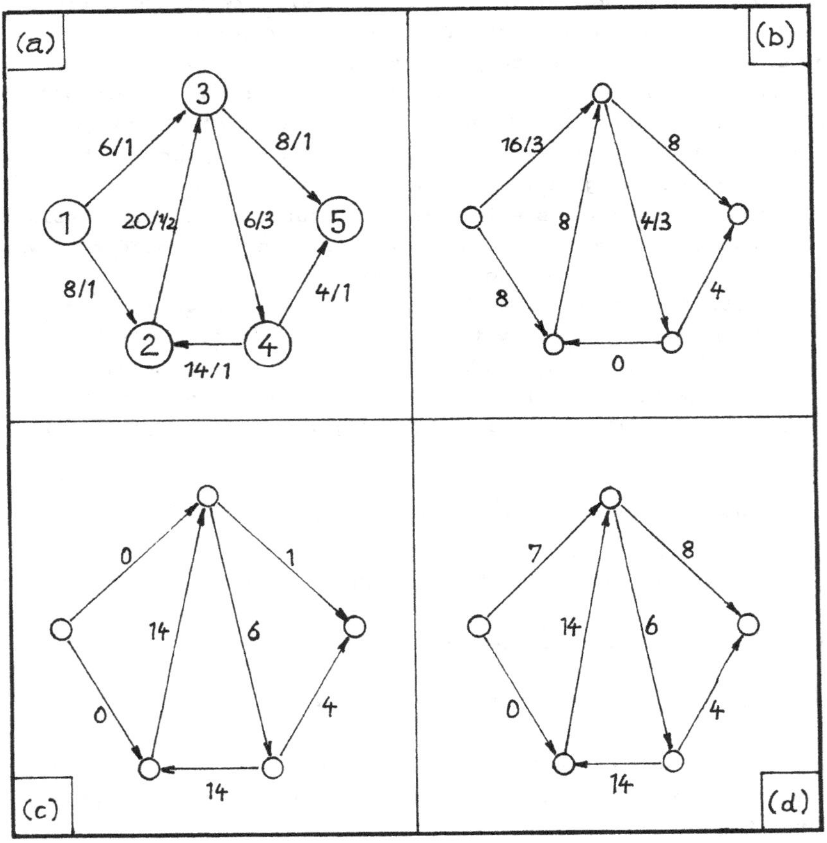

Figure 4.1. (a) Graph G = (V,A). The values along the arcs $(i,j) \in A$ denote cap(i,j) respectively g(i,j).

(b) n-maximum flow x which is not a max-min flow.

(c) max-min flow which is not n-maximum.

(d) n-optimal flow.

The search for n-optimal flows is abbreviated by

MGF opt $\{x_n : x \in \mathbf{X}_g\}$.

The notion of the residual graph $R_g(x) = (V,A(x))$ for a given generalized flow $x \in \mathbf{X}_g$ is the same as in the case of pure flows. Consequently, A(x) is composed of $A^+(x)$ and $A^-(x)$ with

$$A^+(x) = \{(i,j): (i,j) \in A \;\&\; x(i,j) < cap(i,j)\}.$$
$$A^-(x) = \{(i,j): (j,i) \in A \;\&\; x(j,i) > 0\}.$$

With respect to R(x), multipliers are defined as

$$g(i,j) := \begin{cases} g(i,j) & \text{for } (i,j) \in A^+(x) \\ 1/g(i,j) & \text{for } (i,j) \in A^-(x). \end{cases}$$

Let L = [i1,i2,...,il=i1] be an elementary cycle of G. Then

$$g(L) := \prod_{(i,j)\in L+} g(i,j) \Big/ \prod_{(i,j)\in L-} g(i,j)$$

gives the amplification of L, i.e., the amount of flow reaching i1 when one unit of flow is send from i1 and pushed along L. The amplification g(P) for a path P is defined analogously in (3).

A cycle is called *generating* for a given flow $x \in \mathbf{X}_g$ if
(i) $g(L) > 1$ and
(ii) $(i,j) \in L$ implies $(i,j) \in A^+(x)$ for all arcs of L.

The triple (L,k,P) is called *n-generation* if
(i) L is a generating cycle, and
(ii) there is a directed (flow augmenting) path P[k,n] from vertex k of L to n in R(x).

Correspondingly, (L,k,P) is called a *1-generation* if L is a generating cycle and there is a directed path P[k,1] in R(x). A cycle L is called *isolated* for a flow $x \in \mathbf{X}_g$ if there is no vertex k in L such that there is a path P[k,n] or P[k,1] in R(x). For the example of Figure 4.1., the triple (L,k,P) with L = [2,3,4], k = 2, and P = [2,1] is a 1-generation for the flow $x \in \mathbf{X}_g$ shown in (b).

In the case of generalized networks, we can not be certain that every flow unit leaving the source ultimately arrives at the sink because of the existence of absorbing cycles. Additionally, there may be flow units arriving at the sink which need not come from the source because of the existence of generating cycles.

There are two main advantages of pure networks when compared with generalized ones. This concerns firstly the greater efficiency of solution algorithms for pure networks. Secondly, no round-off error is introduced while solving pure network problems. The following two theorems are related to the question to transform a generalized into a pure network problem:

Theorem 4.1. (Truemper 1976)
A generalized flow problem can be converted into a pure flow problem if and only if there exist vertex numbers $p(j)$ for each vertex $j \in V$ such that $p(i) \cdot g(i,j) = g(j)$ for all arcs $(i,j) \in A$.

■

Theorem 4.2. (Glover & Klingman 1973)
Any generalized flow problem whose incidence matrix does not have full row rank can be transformed into a equivalent pure network flow problem.

■

The following results of Onaga (1967) are the foundation of a solution method for MGF:

Theorem 4.3.
A generalized flow $x \in \mathbf{X}_g$ is a max-min flow if and only if each generating cycle L in G(x) is isolated.

■

For $x \in \mathbf{X}_g$ and the residual graph G(x) we use $\mathbf{P}(x)$ for the set of all flow augmenting paths from 1 to n. The weight of a path P is defined as

(3) $g(P) := \prod_{(i,j) \in P+} g(i,j) \;/ \prod_{(i,j) \in P-} g(i,j)$

Theorem 4.4.
Let $x \in \mathbf{X}_g$ be a max-min flow, and let $P \in \mathbf{P}(x)$ be a flow augmenting path of maximum weight, then the flow $x' = x + \alpha \cdot dx$ with sufficiently small $\alpha > 0$ and

$dx(i,j) := char_P(i,j) \cdot g(P[1,i])$ for all $(i,j) \in A$

is a max-min flow again.

■

Corollary:

The flow x' of the above theorem remains feasible as long as $\alpha \leq \min\{\alpha 1, \alpha 2\}$ with

$\alpha 1 := \min\{(cap(i,j) - x(i,j))/g(P[1,i]): (i,j) \in P^+\}$ and

$\alpha 2 := \min\{x(i,j)/g(P[1,i]): (i,j) \in P^-\}$.

■

Theorem 4.5.

A generalized flow $x \in \mathbf{X}_g$ is a solution of MGF if and only if

(i) there exists a cut (X,X^*) such that

$(i,j) \in \delta^+(X)$ implies $x(i,j) = cap(i,j)$

$(i,j) \in \delta^-(X)$ implies $x(i,j) = 0$,

(ii) each generating cycle L in G(x) is isolated.

■

From an algorithmic point of view, the calculation of generating cycles and of paths with maximum multiplicative weight seem to be difficult. However, using the transformation

$$w(i,j) = \begin{cases} -\log g(i,j) & \text{for } (i,j) \in A^+(x) \\ -\log(1/g(i,j)) & \text{for } (i,j) \in A^-(x), \end{cases}$$

both problems can be reduced to well known combinatorial problems:

Lemma 4.1.

Let be $x \in \mathbf{X}_g$ and assume a cycle L in G(x), then

(i) $g(L) > 1$ if and only if $w(L) < 0$

(ii) $g(P') = \max\{g(P): P \in \mathbf{P}(x)\}$ if and only if

$w(P') = \min\{w(P): P \in \mathbf{P}(x)\}$.

Proof:

The propositions are an immediate consequence of the equalities $w(P) = -\log(g(P))$ and $w(L) = -\log(g(L))$.

■

In this way, the calculation of generating cycles and of paths having maximum (multiplicative) weight can be reduced to the calculation of negative cycles respectively shortest paths related to an additive weight. The currently best (from the point of view of computational complexity) strongly polynomial algorithm for the shortest path problem is due to Fredman & Tarjan (1984) who use a Fibonacci heap data structure. The Fibonacci heap takes an average of $O(\log n)$ time for each distance update. This results in an $O(m + n \cdot \log n)$ implementation of the original Dijkstra (1959) algorithm.

4.2. A Combinatorial Algorithm for the Generalized Circulation Problem

Goldberg, Plotkin & Tardos (1988) developed the first polynomial-time combinatorial algorithms for the generalized maximum flow problem, i.e., to maximize the amount of flow excess at the source. We describe the first of their two algorithms based on the repeated application of a minimum-cost flow subroutine.

A *generalized pseudoflow* is a function $w: A \longmapsto R_+$ satisfying the capacity constraints (2). A *generalized circulation* is a generalized pseudoflow w that satisfies conservation constraints

$$\Sigma_{(i,j)\in A}\, g(i,j)\cdot w(i,j) - \Sigma_{(j,k)\in A}\, w(j,k) = 0$$

at all vertices j except at the source 1. A *value* v(w) of a generalized pseudoflow w is defined as

$$v(w) := \Sigma_{(i,1)\in A}\, g(i,1)\cdot w(i,1) - \Sigma_{(1,k)\in A}\, w(1,k).$$

The *generalized circulation problem* is to determine a generalized circulation of highest possible value. It is assumed that the capacities are integers represented in binary and that each gain is given as a ratio of two integers. Then the value of the biggest integer used to represent the gains and capacities is denoted by F. Instead of solving the generalized circulation problem directly, a restricted version with the same input as the original one but with the additional assumption that all flow-generating cycles in the residual graph $R_g(0)$ of the zero flow pass through the source, is considered. The reduction from an instance (G = (V,A),cap, g,1) of the original problem into an instance $(G^* = (V^*,A^*),cap^*,g^*,1)$ of the restricted problem is formalized in procedure REDUCTION.

Lemma 4.2. (Goldberg, Plotkin & Tardos 1988)
h^* is a generalized circulation of maximum value in the restricted problem if and only if the residual graph $R_g(h^*)$ contains no flow-generating cycles. Given h^*, one can construct a maximum value generalized circulation of the original problem in $O(m\cdot n)$ time.

■

One of the crucial points in the combinatorial algorithm is the operation of relabeling. Given a function $\mu: A \longmapsto R_+$ and a generalized network N = (G,cap,g), the *relabeled network* is $N_\mu = (G,cap_\mu,g_\mu)$ with

$cap_{\mu}(i,j) := cap(i,j)/\mu(i)$ and
$g_{\mu}(i,j) := g(i,j)\cdot\mu(i)/\mu(j)$.

Given a generalized pseudoflow w and a labeling μ, the relabeled residual capacity is defined by
$rcap(i,j) := [cap(i,j) - w(i,j)]/\mu(i)$.

```
procedure REDUCTION
begin
  for all (i,j) ∈ A do
C Definition of a generalized pseudoflow h
    if g(i,j) > 1 then h(i,j) := cap(i,j)
                  else h(i,j) := 0
  for all j ∈ V do
    e_g(j) := Σ (i,j)∈A g(i,j)·h(i,j) - Σ (j,k)∈A h(j,k)
    if e_g(j) > 0 then
                  begin
                    A* := A* + {(j,1)}
                    g(j,1) := F^n + 1
                    cap(j,1) := -e_g(j)
                    h(j,1) := 0
                  end
    if e_g(j) < 0 then
                  begin
                    A* := A* + {(1,j)}
                    g(1,j) := F^n + 1
                    cap(1,j) := e_g(j)/g(1,j)
                    h(1,j) := 0
                  end
end
```

If we want to push additional flow from 1, a *canonical relabeling from the source* is performed in the residual network. This relabeling applies when every vertex j is reachable from 1 via a path in the residual network of the actual generalized pseudoflow. For all $k \in V$, $\mu(k) := \max\{g(P): P$ is an elementary path from 1 to k in $R_g(w)\}$.

Theorem 4.6. (Goldberg, Plotkin & Tardos 1988)
After a canonical relabeling from the source:

(i) Every arc (i,j) with non-zero residual capacity, other than the arcs entering the node s, has $g_{\mu}(i,j) \le 1$.

(ii) For every vertex j, there exists a path 1 to k in the residual graph with $g_\mu(i,j) = 1$ for all arcs (i,j) on the path.

(iii) The most efficient flow-generating cycle consist of an $(1,k)$-path for some $k \in V$ with $g_\mu(i,j) = 1$ along the path, and the arc $(k,1) \in R_g(w)$ such that $g_\mu(k,1) = \max \{g_\mu(i,j): (i,j) \in R_g(w)\}$.

■

Theorem 4.7. (Goldberg, Plotkin & Tardos 1988)
A generalized circulation w in a restricted problem is optimal if there exist a labeling μ such that every arc (i,j) in the residual graph $R_g(w)$ has $g_\mu(i,j) \le 1$.

■

The combinatorial algorithm of Goldberg, Plotkin & Tardos (1988) is formalized in procedure COMBINA. It starts with the zero flow. The only iteration that creates a positive excess is the first one with an excess at the source. Each subsequent iteration canonically relabels the residual graph, solves the corresponding minimum-cost flow problem in the relabeled network, and interprets the result as a generalized augmentation. In each iteration, the minimum-cost flow satisfies the deficits that were left after the previous iteration.

```
procedure COMBINA
begin
  w := 0
  repeat
  begin
    for all j ∈ V do
      μ(j) := max {g(P): P is a simple path from 1 to j in Rg(w)}.
      e(j) := Σ (i,j)∈A g(i,j)·w(i,j) - Σ (j,k)∈A w(j,k)
    for all (i,j) ∈ Ag(w) do
      gμ(i,j) := g(i,j)·μ(i)/μ(j)
      c(i,j) := - log gμ(i,j)
      rcap(i,j) := [cap(i,j) - w(i,j)]/μ(i)
    for all j ∈ V do b(j) := - e(j)
C Assume X(rcap,b) to be the flow polyhedron defined by rcap and b
    z := arg min {cTx: x ∈ X(rcap,b)}
    for all (i,j) ∈ Ag(w) do
      if z(i,j) ≥ 0 then dw(i,j) := z(i,j)
                    else dw(i,j) := -g(j,i)·z(j,i)
      w(i,j) := w(i,j) + dw(i,j)·μ(i)
  end
  until gμ(i,j) ≤ 1 for all (i,j) ∈ Ag(w) & e(j) = 0 for all j ∈ V-{1}
end
```

Theorem 4.8. (Goldberg, Plotkin & Tardos 1988)
The above generalized flow algorithm can be implemented so that it will use at most $O((n^2 \cdot \log F) \cdot T(m,n))$ arithmetic operations on numbers whose size is bounded by $O(m \cdot \log F)$, where $T(m,n)$ denotes the complexity for solving the minimum-cost flow problem.

■

4.3. The Simplex Method for Minimum-Cost Generalized Flows

We extend the max flow problem of the previous section by assuming for each arc $(i,j) \in A$ shipping cost per unit of flow, $c(i,j)$. In this model, each vertex is either a supply vector where units of the good enter the network, a demand vector where units leave, or a transshipment vertex. The problem is to minimize total costs with flows that satisfy the capacity constraints and preserve the conservation of flow at each vertex.

Taking $I_g(G)$ for the generalized vertex-arc incidence matrix in which each column contains at most two nonzero coefficients of opposite sign. We will assume that in the column corresponding to an arc $(i,j) \in A$ we have '-1' in row i, coefficient $g(i,j)$ in row j and zero for all remaining elements. Columns with a single nonzero entry are associated with arcs incident on only one vertex, and such arcs are called self-loops. We also assume that $I_g(G)$ has rank n, because otherwise the generalized flow problem can be transformed into a pure one due to Theorem 4.2. The right-hand side vector b associates a requirement $b(j)$ with each vertex $j \in V$.

GMCF $\min \{c^T x: I_g(G) \cdot x = b,\ 0 \le x \le cap\}$.

To formulate the dual problem $GMCF^d$ we associate dual variables $p(i)$ for all vertices $i \in V$ in correspondence to the flow conservation rules (1), and dual variables $y(i,j)$ in correspondence to the upper capacity constraints (2) of all $(i,j) \in A$.

GMCF^d $\max \{p^T b - y^T cap :$

$$g(i,j) \cdot p(j) - p(i) - y(i,j) \le c(i,j) \quad \text{for all } (i,j) \in A$$

$$y(i,j) \ge 0 \quad \text{for all } (i,j) \in A\}$$

We define $tg(i,j) := g(i,j) \cdot p(j) - p(i)$ and use

$$y(i,j) := \max \{0, tg(i,j) - c(i,j)\} \quad \text{for all } (i,j) \in A.$$

Theorem 4.9.
(x,p) is a primal and dual optimal solution pair if and only if

(4) $x(i,j) = 0$ for all $(i,j) \in A$ with $tg(i,j) < c(i,j)$

(5) $0 \le x(i,j) \le cap(i,j)$ for all $(i,j) \in A$ with $tg(i,j) = c(i,j)$

(6) $x(i,j) = cap(i,j)$ for all $(i,j) \in A$ with $tg(i,j) > c(i,j)$

(7) $d(j) = b(j) + \Sigma_{(j,k)\in A}\, x(j,k) - \Sigma_{(i,j)\in A}\, g(i,j)\cdot x(i,j) = 0$ for all $j \in V$.

Proof:
From the optimality conditions of linear programming we obtain

$y(i,j)[cap(i,j)-x(i,j)] = 0$ for all $(i,j) \in A$

$x(i,j)[c(i,j) + p(i) - g(i,j)\cdot p(j) + y(i,j)] = 0$ for all $(i,j) \in A$

The case $t_g(i,j) < c(i,j)$ implies $x(i,j) = 0$ and $t_g(i,j) > c(i,j)$ results in $y(i,j) > 0$. Consequently, it must be $x(i,j) = cap(i,j)$. (5) and (7) ensures the primal feasibility of x; the dual feasibility is valid using the above definition of y.

■

Again, there are different approaches for achieving optimality. One of them is the specialization of the primal simplex method to generalized networks. A basis B for GMCF is a full row rank matrix composed of a linearly independent set of column vectors of $I_g(G)$. The graph-theoretical interpretation is a collection of quasi-trees (also called one-trees) where a *quasi-tree* is a tree to which a single arc has been added. It is possible for this loop to be incident to just a single vertex (self-loop).

Theorem 4.10. (Dantzig 1963)
Any basis B extracted from a generalized network flow problem can be put in the form (8) by rearranging rows and columns:

$$(8) \qquad B = \begin{bmatrix} B^1 & & & \\ & B^2 & & \\ & & \ddots & \\ & & & B^p \end{bmatrix},$$

where each square submatrix B^q is either upper triangular or nearly upper triangular with only one element below the diagonal.

■

The efficiency of the simplex method for generalized flows is closely related to extremely fast solution updates based on sophisticated data structures for storing each new basis. So it is possible to find the representation of an entering vector and to determine updated values for dual variables by tracing appropriate segments of the quasi-trees. Therefore, no explicit basis inverse is needed, and the usual arithmetic operations required to update the inverse are eliminated.

Figure 4.2 depicts the structure of the basis of a generalized network both in matrix and graph-theoretical representation.

(a)

	(1,3)	(1,2)	(2,3)	(2,4)	(4,5)	(4,6)	(1,7)	(8,8)	(8,9)	(8,10)	(11,12)	(11,13)
1	■	■					■					
2		■	■	■								
3	■		■									
4				■	■	■						
5					■							
6						■						
7							■					
8								■	■	■		
9									■			
10										■		
11											■	■
12											■	■

(b)

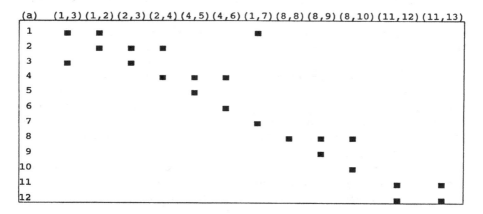

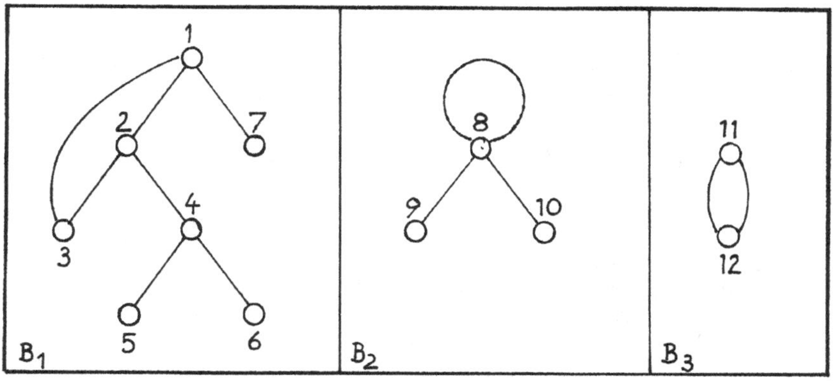

Figure 4.2. Basis structure for generalized networks.
(a) Matrix representation.
(b) Graph-theoretical representation.

Suppose that a quasi-tree Q is given. We select one arc of the loop of Q and designate it as the special arc of Q. One of the end-points of this special arc is taken as the root of the remaining subgraph T which is a tree. The root is regarded pictorially as the highest vertex of T with all other vertices hanging below it. With an arc $(i,j) \in T$ such that i is closer to the root than j, then i is called the *predecessor* of j and j is an *immediate successor* of i. The subtree of T that contains a vertex i and all of its successors will be denoted as T(i). For vertices other than the root, p(i) denotes the predecessor of i. If i is the root, then p(i) is the vertex at the opposite end of the special arc. The preorder successor of vertex i is denoted as s(i). Once a preorder is given, the thread is defined as the function which traces the tree vertices in preorder. The value of the thread of the last vertex in the preorder is defined to be the root. The function whose value at vertex i is the number of vertices in T(i) is denoted by t(i). The last vertex of T(i) in the preorder defining s is denoted by f(i). For vertices on the loop the predecessors are flagged by means of a negative index. All the functions are illustrated in Figure 4.3. by the one-tree B_1 shown in Figure 4.2.(b).

The principal steps of the simplex method for generalized networks are summarized in procedure GN-SIMPLEX. We assume a basis B having a set $A_B \subset A$ of basic arcs. B can be obtained by introducing dummy arcs at the beginning. The actual flow is denoted by x.

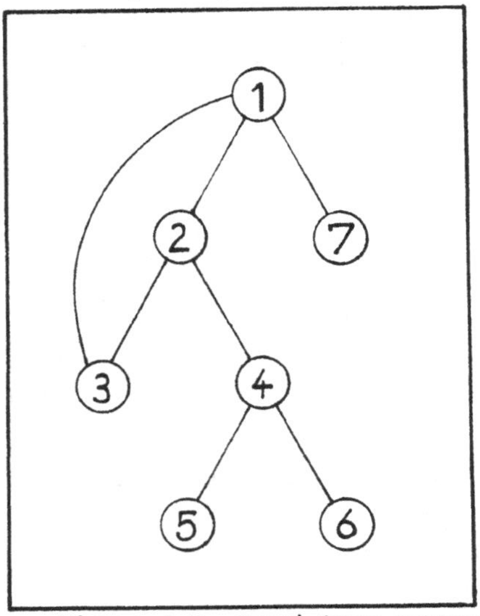

vertex	p	s	f	t
1	-3	2	7	7
2	-1	3	6	5
3	-2	4	3	1
4	2	5	6	3
5	4	6	5	1
6	4	7	6	1
7	1	1	7	1

Figure 4.3. Quasi-tree representation of B_1.

```
procedure GN-SIMPLEX
begin
  repeat
  begin
C cB denotes the costs associated with the set of basic arcs
    p := cB·B^-1     (dual price vector)
    for all (i,j) ∈ A - AB do  rc(i,j) := c(i,j) + g(i,j)·p(i) - p(j)
    AL := {(i,j) ∈ A - AB: x(i,j) = 0 & rc(i,j) < 0}
    AU := {(i,j) ∈ A - AB: x(i,j) = cap(i,j) & rc(i,j) > 0}
    choose (k,l) ∈ AL + AU
    if (k,l) ∈ AL then sgn := 1
                  else sgn := -1
C N[k,l] denotes the column of Ig(G) w.r.t. (k,l) ∈ A - AB
    y := B^-1·N[k,l]
    for h = 1 to n do
                      ⎧  1      if  y(h) > 0
              σ(h) := ⎨ -1      if  y(h) < 0
                      ⎩  0      otherwise
    d1 := min {min {x(ih,jh)/abs(y(h)): (ih,jh) ∈ AB & σ(h) = sgn},∞}
    d2 := min {min {[cap(ih,jh) - x(ih,jh)/abs(y(h))]:  (ih,jh) ∈ AB
                                                     & -σ(h) = sgn},∞}
    delta := min {d1,d2,cap(k,l)}
    for all (ih,jh) ∈ AB do x(ih,jh) := x(ih,jh) - delta·sgn·y(h)
    x(k,l) := x(k,l) + delta·sgn
C Assume that delta is determined by (s,t)
    AB := AB + {(k,l)} - {(s,t)}
  end
  until AL + AU = φ
  if AB contains an artificial variable having flow > 0
  then  the problem has no feasible solution
  else  x is an optimal solution
end
```

An efficient implementation of GN-SIMPLEX can be found in Brown & McBride (1985). They used the so-called preorder traversal method for representing the network basis. Their description includes a more detailed presentation how to calculate the dual multipliers p and how to perform the column update $y = B^{-1} \cdot N[k,l]$. As another solution method, Bertsekas & Tseng (1988) extended their relaxation approach to solve GMCF.

We illustrate the simplex method for generalized flows by an example. For that reason consider the graph shown in Figure 4.4. with the corresponding values of the capacities, multipliers, and costs. The demand-supply vector b is b = (-7,0,0,0,12).

(i,j)	cap(i,j)	g(i,j)	c(i,j)
(1,2)	8	1	3
(1,3)	8	1	2
(1,5)	7	1.5	100
(2,3)	20	0.5	1
(3,4)	6	3	3
(3,5)	8	1	4
(4,2)	14	1	3
(4,5)	4	1	7

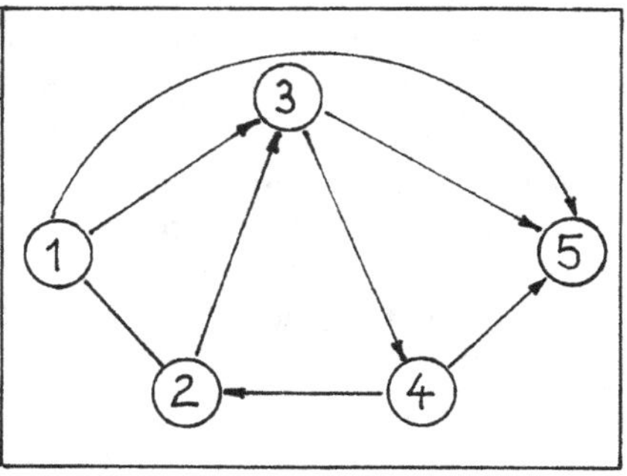

Figure 4.4. Graph with capacities, multipliers, and costs.

As a first basic solution we consider the quasi-tree of Figure 4.5. with flow values given on the arcs. The flow on all non-basic arcs is equal to zero.

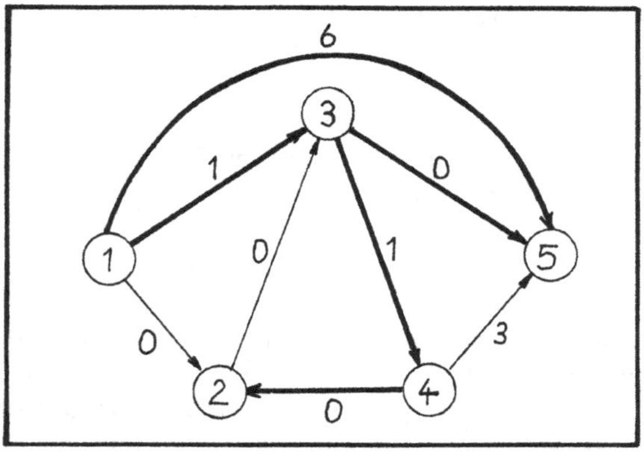

Figure 4.5. Quasi-tree (bold) and flow of the first basic solution.

Taking the near-triangular basis

$$B = \begin{array}{c} \\ 1 \\ 3 \\ 4 \\ 5 \\ 2 \end{array} \begin{array}{c} \begin{array}{ccccc} (1,5) & (1,3) & (3,4) & (4,5) & (4,2) \end{array} \\ \left[\begin{array}{ccccc} -1 & -1 & & & \\ & 1 & -1 & & \\ & & 3 & -1 & -1 \\ 1.5 & & & 1 & \\ & & & & 1 \end{array}\right] \end{array}$$

the dual variables p = (p(1),...,p(5)) are calculated from the equation $p \cdot B = c_B$, where c_B is the cost vector in correspondence to the basic variables. The solution of the system

$$\begin{array}{rrrrrl} -p(1) & & & & +1.5p(5) & = 100 \\ -p(1) & & +p(3) & & & = 2 \\ & & -p(3) & +3p(4) & & = 3 \\ & & & -p(4) & +p(5) & = 7 \\ & p(2) & & -p(4) & & = 3 \end{array}$$

is p = (-174, -53 1/3, -172, -56 1/3, -49 1/3).

The reduced costs for all non-basic arcs are:

rc(1,2) = -117 2/3 < 0
rc(2,3) = 35 2/3 > 0
rc(3,5) = -118 2/3 < 0 .

We take the most violating arc (3,5) for entering the basis. The leaving arc is (4,5). The resulting quasi-tree with corresponding flow values is shown in Figure 4.6. The vector of dual variables is p = (182, 65 1/3, 184, 62 1/3, 188). The flow along the non-basic arc (4,5) is at its upper bound.

The optimality test is not fulfilled for the non-basic arc (2,3), since the reduced costs are rc(2,3) = 1 + 65 1/3 - 92 = -25 1/3 < 0. After one more basis exchange step replacing (3,4) by (2,3) we obtain the quasi-tree of Figure 4.7. The vector p of dual variables is p = (182, 91, 184, 88, 188).

The reduced costs are rc(1,2) = 94 > 0,
rc(3,5) = -77 < 0 for x(3,5) = cap(3,5), and
rc(4,5) = -93 < 0 for x(4,5) = cap(4,5) > 0.

Consequently, the optimality conditions as stated in Theorem 4.9. are satisfied.

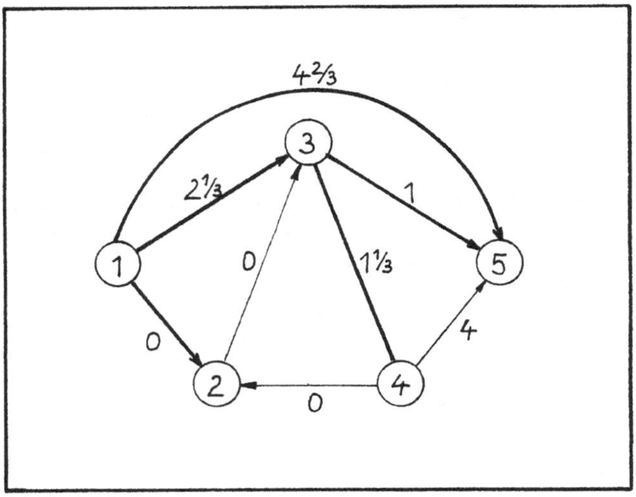

Figure 4.6. Quasi-tree (bold) of the second basic solution.

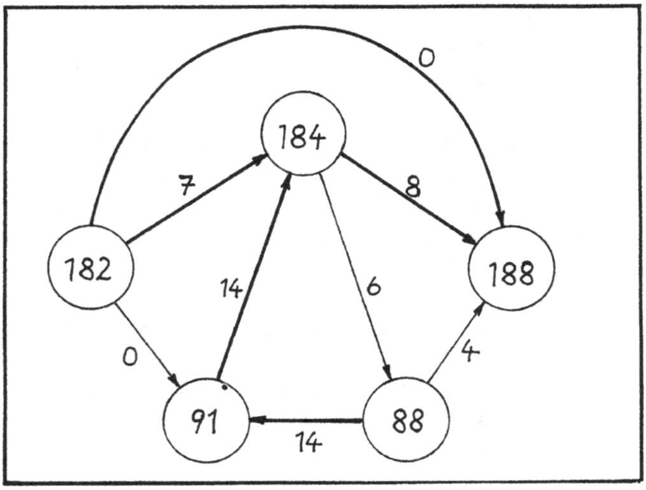

Figure 4.7. Optimal solution.

4.4. Computational Results

Kennington & Muthukrishnan (1988) have developed a generalized network code called GENFLO. They compared GENFLO with two other codes for solving generalized network problems:

- MPSX (the IBM general mathematical programming system),
- GENNET (Brown & McBride 1984).

GENNET assumes that every column of $I_g(G)$ has at most two nonzero entries and one must be one (minus one). GENFLO assumes that every column has at most two nonzero entries and MPSX makes no assumptions about the entries in $I_g(G)$. The test problems for the comparison were generated by a modification of NETGEN called GNETGEN. Both specialized codes were written in Fortran. The main result is a cost saving of ten to one when the specialized codes are used instead of the generalized code MPSX. The greater flexibility with an arbitrary second multiplier allowed in GENFLO results in a computational expense of approximately ten percent. The detailed computer times in seconds on an IBM 3081D are presented in Table 4.1.

Table 4.1. Solution times (in seconds) for generalized networks due to Kennington & Muthukrishnan (1988).

Problem Number	Size #(V)	#(A)	MPSX	GENNET	GENFLO
1	200	1500	7.80	0.62	0.95
2	200	4000	3.00	0.22	0.23
3	200	6000	18.60	2.07	1.53
4	300	4000	47.40	3.50	4.23
5	400	5000	26.20	2.06	2.23
6	400	7000	16.80	1.59	1.68
7	1000	6000	40.20	3.30	3.60

§5 MULTICRITERIA FLOWS

5.1. Fundamental Results

The majority of problems appearing in practice cannot be analyzed adequately without taking into account more than one criterion. We consider a feasible area $\mathcal{B}$ and a vector-valued function $f(x) = (f_1(x),\ldots,f_Q(x))$. In the following, we always consider linear functions $f_k(x) = c_k^T x$ for all $k = 1,\ldots,Q$. These functions are arranged as the Q rows of a matrix C.

The criteria are conflicting in the most cases. That means, with x_j^* denoting the optimal solution of the j-th criteria regardless of the remaining ones; in general it is $x_j^* \neq x_k^*$ for $j,k \in \{1,\ldots,Q\}$ and $j \neq k$. A feasible solution $x \in \mathcal{B}$ of a multicriteria optimization problem is called an *efficient (minimal) solution* (also: non-dominated or Pareto-optimal solution) with respect to f(x), if

(1) there exists no feasible z with $f_k(z) \le f_k(x)$; $k = 1,\ldots,Q$ and $f(z) \neq f(x)$.

For each efficient solution there is no other feasible solution that will improve one criterion without degrading at least one of the others. The set of all efficient solutions has an inner stability in the sense that any two of their members are not comparable. The search for efficient minimal solutions is abbreviated by min*. To summarize some important fundamental results, we consider the general linear programming problem

MCLP $\min^* \{Cx: x \in \mathcal{B}\}$.

The *ideal vector* z^* is defined due to $z_j^* := c_j^T x_j^*$ for $j = 1,\ldots,Q$. The *components of the nadir vector* $w^* \in \mathbf{R}^Q$ are the maxima $w_j^* := \max \{c_j^T x_i^*: 1 \le i \le Q\}$, which is not necessarily unique because some objective functions may have alternative minimum solutions. The ideal vector is usually infeasible, otherwise there would be no conflicting objectives. In many applications the nadir vector is taken as a pessimistic estimate for the outcome of the respective objective functions when decision makers balance their preferences.

The most common strategy to characterize nondominated solutions is in terms of optimal solutions of some scalar optimization problem.

Theorem 5.1.
The following three statements are equivalent:

(i) x1 is an efficient solution of MCLP;

(ii) there is a vector $t \in \mathbf{R}^Q$ with $t_k > 0$ for all $k = 1,\dots,Q$ such that $t^T Cx1 < t^T Cx$ with respect to all feasible $x \in \mathcal{X}$;

(iii) for $k = 1,\dots,Q$ it holds that
$x1 \in \arg\min \{c_k^T x \ : x \in \mathcal{X},\ c_i^T x \le l_i$ for all $i \neq k\}$.
with $l_i = c_i^T x1$.

■

Corollary

(i) If z is a unique solution of $\min \{t^T Cx: x \in \mathcal{X}\}$ for any parameter vector $t \in \mathbf{R}^Q$, $t \geq 0$ then z is an efficient solution.

(ii) If z is a unique solution of
$\min \{c_k^T x: x \in \mathcal{X},\ c_j^T x \le u_j\ ;\ j = 1,\dots,Q,\ j \neq k\}$
for an index $k \in \{1,\dots,Q\}$ and arbitrary upper bound vector u then z is an efficient solution.

■

The above theorem is also valid for general functions f(x) and implies solution methods based on parametric programming (compare Lootsma 1988). However, without convexity assumption, some of the nondominated solutions may never be discovered by scalarization. This is especially true in the case of discrete problems.

Theorem 5.2. (Yu & Zeleny 1975)
The set of all efficient basic solutions of MCLP is connected.

■

From Theorem 5.2. an 'easy' solution method for MCLP can be concluded: Firstly, determine an initial efficient basic solution from a weighted linear combination of the Q linear objective functions. After that, the complete set of efficient basic solutions can be obtained by exhaustively examining all bases adjacent to the set of currently known bases.

Theorem 5.3. (Yu & Zeleny 1975)
Let x be an efficient basic solution and let z be an adjacent basic solution with a Q x n - matrix of dual variables given by $\cap = C_B(B_z)^{-1}$ where C_B is the Q x m - submatrix of C corresponding to the basic variables and $(B_z)^{-1}$ is the inverse of the basis B_z in correspondence to z. Then z is an efficient solution iff

(2) $\max \{1^T e : I \cdot e + (C - \cap \cdot A)y = 0,\ e \geq 0,\ y \geq 0\} = 0.$

■

The non-dominance subproblem (2) is relatively easy to solve since an initial feasible basis is readily available. However, this approach needs the solution of (2) for each of the adjacent bases of the given efficient basic solution x. Further computational aspects of this question are discussed in connection with the multicriteria network simplex method in Section 5.3.

To avoid efficient solutions for which the marginal improvement in one objective function value related to the losses in all other components is unbounded from above, Geoffrion (1968) introduced the notion of *proper efficiency*. In the linear case as considered here, all efficient solutions are also properly efficient.

5.2. Complexity Results

Despite the fact that multicriteria combinatorial optimization problems arise quite often in real world problems, it needed a quite long time till Serafini (1987) presented the computational complexity of this class of problems in precise terms. Before presenting some detailed results related to multicriteria flow problems, we give a list of several versions presented by Serafini to 'solve' a multicriteria combinatorial optimization problem. Given a vector-valued function $f: Z^n \longmapsto Z^Q$, 'find' minimal solutions in a subset $F \subset Z^n$.

The minimality is used as defined in (1). For any instance (f,F), the following (and some more) questions may be investigated:

V_1: find all minimal solutions;

V_2: find all minimal values and at least one minimal solution for each minimal value;

V_3: find all minimal values;

V_4: find at least $p > 1$ minimal values or, in case p minimal values do not exist, report this fact and find all minimal values. Furthermore, find at least one minimal solution for each minimal value;

V_5: find at least $p > 1$ minimal values or, in case p minimal values do not exist, report this fact and find all minimal values;

V_6: find one minimal solution;

V_7: find one minimal value;

V_8: given $z \in Z^Q$, does there exist $x \in F$: $f(x) \leq z$?

V_9: given $z, w \in Z^Q$, $w > 0$, does there exist $x \in F$: $w^T f(x) \leq w^T z$?

With the exception of V_1, all versions can be reduced to V_8. As a conclusion, V_8 is suggested to be a standard reference version to measure computational complexity. In this sense, the NP-completeness of the restricted shortest path problem is of fundamental importance:

Let $c,d: A \longmapsto \mathbb{R}$ be two cost functions defined on the set of arcs A of a directed graph G.

Theorem 5.4. (Garey & Johnson 1979)
Let δ_1, δ_2 be two given constants. Then the following problem is NP-complete: Decide whether there is a directed path P from 1 to n in G such that

$$c(P) := \Sigma_{(i,j)\in P}\, c(i,j) \le \delta_1 \text{ and}$$
$$d(P) := \Sigma_{(i,j)\in P}\, d(i,j) \le \delta_2.$$

■

From this theorem it follows that all multicriteria flow problems containing the decision version of the bicriteria shortest path problem as a special case are NP-complete when formulated in terms of V_8. Since flow problems can be handled using linear programming, we take two other characterizations of complexity into consideration. For an arbitrary multicriteria problem MCP we denote by Φ_1(MCP) the number of efficient solutions such any two of these solutions are not mapped into the same objective function vector. Φ_2(MCP) expresses the number of efficient extreme point solutions in the objective space. As an immediate consequence it follows that $\Phi_1(\text{MCP}) \ge \Phi_2(\text{MCP})$. The results to determine all non-dominated extreme point solutions are not good, except for two objective problems (Zionts 1985). In Ruhe (1988b) it was shown that for network flow problems it is sufficient to consider $Q = 2$ objective functions to prove an exponential number of efficient extreme points in the objective space. The bicriteria generalization of the minimum-cost flow problem is:

BMCF $\min^* \{(c^Tx, d^Tx): x \in \mathbf{X}\}$.

As a special case, for $\#(x) = 1$ we obtain the bicriteria shortest path problem.

Lemma 5.1. (Hansen 1980)
The bicriteria shortest path problem is, in the worst case, intractable, i.e., requires for some problems a number of operations growing exponentially with the problems input size.

■

In the pathological instance proving Lemma 5.1., the minimal points in the objective space are located on a line. We investigate the question , whether for BMCF there is also an exponential number Φ_2(BMCF) of efficient solutions which correspond to extreme points in the objective space. For that reason, a graph originally introduced by Zadeh (1973) for demonstrating the pathological behaviour of various algorithms for the minimum-cost flow problem, was taken into account. The graph $G_n = (V_n, A_n)$ with $\#(V_n) = 2n + 2$ and with $\#(A_n) = O(n^2)$ arcs is shown in Figure 5.1. G_n is acyclic, i.e., it does not contain a directed cycle. The pathological behaviour for the minimum-cost flow problem is caused by a special sequence of flow augmenting paths. These paths are shown in Figure 5.2. for the graph G_3.

The augmentations are based on the following upper capacity bounds defined for the arcs of G_n:

$$(3)\quad cap(i,j) = \begin{cases} 1 & \text{for } (i,j) = (0,1) \\ 2 & \text{for } (i,j) = (\underline{1},\underline{0}), (\underline{2},\underline{0}) \\ 3 & \text{for } (i,j) = (0,2) \\ 2^{k-1} + 2^{k-3} & \text{for } (i,j) = (0,k), (\underline{k},\underline{0});\ k = 3,\ldots,n \\ \infty & \text{otherwise} \end{cases}$$

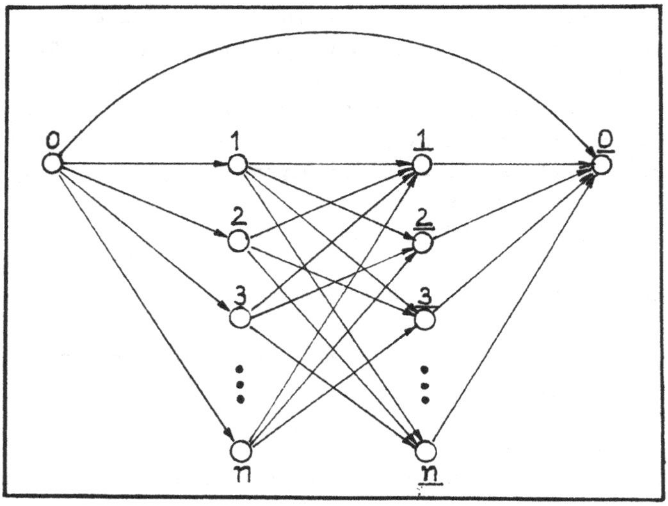

Figure 5.1. The Zadeh graph $G_n = (V_n, A_n)$.

The sequence of augmentations can be described recursively. For n = 2, the paths are:

$P_1 = [0,1,\underline{1},\underline{0}]$
$P_2 = [0,2,\underline{1},\underline{0}]$
$P_3 = [0,2,\underline{1},1,\underline{2},\underline{0}]$.

Let $P_1,\ldots,P_{r(k)}$ be the paths in G_k, then there are three types of paths in G_{k+1}:

(4) the paths $P_1,\ldots,P_{r(k)}$ of G_k,
(5) $P_{r(k)+1} = [0,k,\underline{k+1},\underline{0}]$,
$P_{r(k)+2} = [0,k+1,\underline{k},\underline{0}]$, and
(6) a sequence of paths of the form
$P_{r(k)+2+h} = [0,k+1,\overleftarrow{P}_{r(k)+1-h},\underline{k+1},\underline{0}]$; $h = 1,\ldots,r(k)$
where $\overleftarrow{P}_l$; $l = 1,\ldots,r(k)$ denotes the path which results from P_l by omitting the arcs incident to 0 or $\underline{0}$ and by reversing the direction of P_l.

Let Γ: $A \longmapsto R$ be a weight function, then for each path P

$$\Gamma(P) := \Sigma_{(i,j)\in P^+} \Gamma(i,j) - \Sigma_{(i,j)\in P^-} \Gamma(i,j)$$

is defined. In the following, we always use:

$$(7)\quad \Gamma(i,j) := \begin{cases} \infty & \text{for } (i,j) = (0,\underline{0}) \\ 0 & \text{for } (i,j) = (0,k),(\underline{k},\underline{0});\ k=1,\ldots,n \\ 2^{\max\{i,j\}-1} - 1 & \text{otherwise} \end{cases} .$$

Lemma 5.2. (Zadeh 1973)
Assume the graph G_n and the weight function as defined in Figure 5.1. respectively in (7). Then the following propositions are valid:

(i) The recursive application of (4) - (6) results in $r(n) = 2^n + 2^{n-2} - 2$ flow augmenting paths. Each of these augmentations increases the total flow in the network by 1.
(ii) For $1 \le k \le r(n)-1$, it holds $\Gamma(P_k) \le \Gamma(P_{k+1})$ with the additional property that $\Gamma(P_k) = \Gamma(P_{k+1})$ implies $\#(P_k) = \#(P_{k+1})$.
(iii) Each augmentation sends flow along exactly one arc of the form $(i,\underline{j})$ which did not carry flow in the preceding iteration.
(iv) Each augmentation either increases the flow in exactly one arc to its capacity or decreases the flow in exactly one arc to zero. ∎

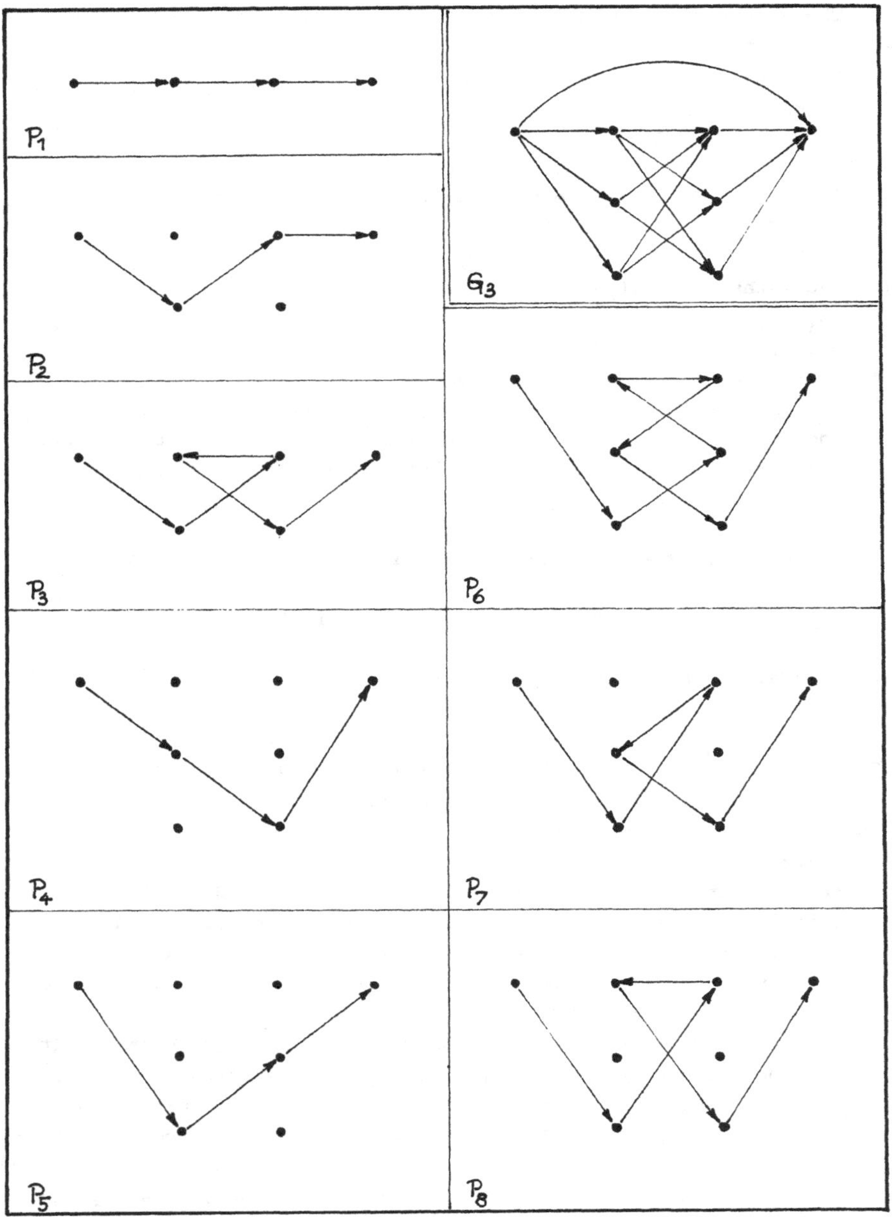

Figure 5.2. The flow augmenting paths $P_1, P_2, \ldots, P_8$ in G_3.

Lemma 5.3.
Among the r(n) paths $P_1,..,.P_{r(n)}$ in G_n there are exactly $s(n) = 2^n - 1$ paths with index set I(n) such that

(8) $\Gamma(P_r) < \Gamma(P_s)$ for all $r,s \in I(n)$ with $r < s$.

Proof:
The proposition is shown by proving that there are $w(n) = 2^{n-2} - 1$ pairs of paths in G_n having the same weight. For G_3 we obtain w(3) = 1. The only pair is P_4, P_5. Assume that $w(k) = 2^{k-2} - 1$ and consider n = k+1, then the number w(k+1) of pairs in G_{k+1} can be determined by the application of (4) - (6). From (4), w(k) pairs are known. (5) contributes to one more pair. Finally, (6) gives once more w(k) pairs. Consequently, $w(k+1) = 2 \cdot w(k) + 1 = 2^{k-1} - 1$. There remains s(n) = $r(n) - w(n) = 2^n - 1$ paths satisfying (8).

■

To demonstrate that BMCF may have an exponential number (in dependence of n) of extreme points in the objective space we define a pathological instance based on G_n.

$$(9) \quad c(i,j) = \begin{cases} 1 & \text{for } (i,j) = (0,k),(\underline{k},\underline{0});\ k=1,\dots,n \\ & \text{and } (i,j) = (0,\underline{0}) \\ 2^{\max\{i,j\} - 1} - 1 & \text{otherwise.} \end{cases}$$

$$(10) \quad d(i,j) = \begin{cases} 1 & \text{for } (i,j) = (\underline{k},\underline{0});\ k = 1,\dots,n \\ 2 & \text{for } (i,j) = (0,\underline{0}) \\ 0 & \text{otherwise.} \end{cases}$$

Additionally, we use the capacity bounds as given in (3). Consequently, each of the paths $P_1,\dots,P_{r(k)}$ causes a maximal flow augmentation which increases the total flow by 1.

Theorem 5.5.
The instance of BMCF defined by G_n,(3),(9),(10), and $K = 2^n + 2^{n-2} - 2$ has Φ_1(BMCF) = K + 1 efficient basic solutions $x_0,\dots,x_K$ with

$$f(x_0) = (K,2K),$$

$$(11) \quad f(x_p) = (K + p + \sum_{k=1}^{p} \Gamma(P_k), 2K - p) \text{ for } p = 1,\dots,K.$$

Among them are exactly Φ_2(BMCF) = 2^n extreme points in the objective space.

Proof:
Consider the paths $P_1,\dots,P_{r(n)}$ as given above and $P_0 = [0,\underline{0}]$. The

flows are defined as the linear combination of augmentations along these paths. Let $z = (z(0), z(1), \ldots, z(r(n))) \in R^{r(n)+1}$, then we consider flows of the form

$$(12) \quad x = \sum_{k=0}^{r(n)} \text{char}(P_k) \cdot z(k).$$

Since BMCF is totally unimodular and all capacity bounds are integer it follows that $z(k) \in \{0,1\}$ for all $k = 1, \ldots, r(n)$. As a consequence of $\#(x) = K$ $(= r(n))$ we obtain that

$$z(0) = K - \sum_{k=1}^{r(n)} z(k) .$$

Let $x_0, x_1, \ldots, x_{r(n)}$ be a sequence of flows defined by $z_0 = (K, 0, \ldots, 0)$ and z_p; $p = 1, \ldots, r(n)$:

$$z_p(k) = \begin{cases} K - p & \text{for } k = 0 \\ 1 & \text{for } 1 \le k \le p \\ 0 & \text{otherwise.} \end{cases} \quad ; \; p = 1, \ldots r(n)$$

With $f(x) = (c^T x, d^T x)$ the corresponding vectors of objective function values are:

$$f(x_0) = (K, 2K)$$

$$f(x_p) = (K + p + \sum_{k=1}^{p} \Gamma(P_k),\; 2K - p);\; p = 1, \ldots, r(n).$$

For the spanning tree T_0 related to x_0 we refer to Figure 3.3. Since $x_0(i,j) = 0$ for each $(i,j) \in A - T_0$ and $0 \le x(i,j) \le cap(i,j)$ for each $(i,j) \in T_0$, x_0 is a basic solution corresponding to T_0. The spanning trees $T_1, \ldots, T_{r(n)}$ related to $x_1, \ldots, x_{r(n)}$ are defined recursively. According to (iii) and (iv) of Lemma 5.2., for each flow augmenting path P_k; $k = 1, \ldots, r(n)$, there is exactly one arc $(i_k, j_k) \in P_k$ such that

$$\sum_{h=0}^{k-1} x_h(i_k, j_k) = 0 \quad \text{and} \quad x_k(i_k, j_k) > 0.$$

Additionally, there is always an arc $(i_k', j_k') \in T_{k-1}$ such that either

$$\sum_{h=0}^{k-1} x_h(i_k', j_k') > 0 \quad \text{and} \quad \sum_{h=0}^{k} x_h(i_k', j_k') = 0 \quad \text{or}$$

$$\sum_{h=0}^{k-1} x_h(i_k',j_k') < cap(i_k',j_k') \text{ and } \sum_{h=0}^{k} x_h(i_k',j_k') = cap(i_k',j_k').$$

Using $T_k := (T_{k-1} - \{(i_k',j_k')\}) + \{(i_k,j_k)\}$, from x_{k-1}, T_{k-1} a new basic solution pair x_k,T_k with a spanning tree T_k is obtained.

To prove that the solutions x_l; $l = 0,\ldots,r(n)$ are indeed efficient solutions, the equivalence between multiobjective and parametric programming as stated in Theorem 5.1. is used. For BMCF that means: x is an efficient (basic) solution if and only if there is a parameter $\mu \in (0,1)$ such that x is a solution of:

(13) $\min \{(1-\mu)\cdot c^T x + \mu\cdot d^T x\colon x \in \mathbf{X}\}$

Due to the form (12) of the flows and $(c(P_0),d(P_0)) = (1,2)$, $(c(P_1),d(P_1)) = (2,1)$, and $(c(P_k),d(P_k)) \geq (2,1)$ for each $k = 2,\ldots,r(n)$ we confirm the optimality of x_0 for $\mu \in (0,1/2]$ since $1 - \mu + 2\cdot\mu \leq 2(1 - \mu) + \mu$. A path P with $(c(P),d(P)) = (k,1)$ is preferred to P_0 if $(1 - \mu)\cdot k + \mu < 1 - \mu + 2\cdot\mu$, i.e., $\mu > (k-1)/k$. In this way, the intervals $(0,1/2]$, $[1/2,2/3],\ldots$, $[(2^n-1/2^n),1)$ are determined. Each of the basic solutions $x_0,\ldots,x_{r(n)}$ is optimal in one of the given intervals, i.e., it is an efficient solution. This is the first part of the proposition.

Between the vectors of consecutive objective function values we define

$$\begin{aligned} \text{slope}(p) &:= d^T(x_{p+1} - x_p)/c^T(x_{p+1} - x_p) \\ &= 1/(1 + \Gamma(P_{p+1}))\ ; \qquad p = 1,\ldots,r(n). \end{aligned}$$

With the $s(n) = 2^n - 1$ paths described in Lemma 5.3. we obtain $2^n - 1$ different slopes, or equivalently, 2^n efficient extreme points in the objective space.

■

To illustrate the above considerations, the characteristics of the pathological instance of Theorem 5.5. for the graph G_3 are summarized in Table 5.1. The right column gives the stability interval with respect to (13) for each optimal solution x_p.

Table 5.1.

k	$\Gamma(P_k)$	$(c(P_k),d(P_k))$	p	$z_p = (z_p(0),...,z_p(8))$	$f(x_p)$	stab.int.
0	∞	(1,2)	0	(8,0,0,0,0,0,0,0,0)	(8,16)	(0,1/2]
1	0	(2,1)	1	(7,1,0,0,0,0,0,0,0)	(9,15)	[1/2,2/3]
2	1	(3,1)	2	(6,1,1,0,0,0,0,0,0)	(11,14)	[2/3,3/4]
3	2	(4,1)	3	(5,1,1,1,0,0,0,0,0)	(14,13)	[3/4,4/5]
4	3	(5,1)	4	(4,1,1,1,1,0,0,0,0)	(18,12)	[4/5,5/6]
5	3	(5,1)	5	(3,1,1,1,1,1,0,0,0)	(22,11)	[5/6,5/6]
6	4	(6,1)	6	(2,1,1,1,1,1,1,0,0)	(27,10)	[5/6,6/7]
7	5	(7,1)	7	(1,1,1,1,1,1,1,1,0)	(33, 9)	[6/7,7/8]
8	6	(8,1)	8	(0,1,1,1,1,1,1,1,1)	(40, 8)	[7/8,1)

5.3. Algorithms

5.3.1. Lexicographic Optimization

In lexicographic optimization, the objective functions are assumed to be arranged according to decreasing importance. Then a low priority objective is only optimized as far as it does not interfere with the optimization of higher priority objectives.

Let v,w be two vectors from $\mathbf{R}^m$. v is called *lexicographically less* than w if $v(i) = w(i)$ for $i = 1,...,j-1$ and $v(j) < w(j)$ for $j = \min\{k: 1 \le k \le m, v(k) \neq w(k)\}$. This is written as $v \ll w$. The search for the lexicographically smallest element is denoted by 'lex min'. To demonstrate the relationship between multicriteria and lexicographic optimization, we investigate the generalization of the minimum-cost flow problem MCF to the problem LMCF with lexicographic objective function vector

$$\textbf{LMCF} \quad \text{lex min } \{(c_1^Tx,...,c_Q^Tx) : x \in \mathbf{X}\}$$

and the multicriteria minimum-cost flow problem

$$\textbf{MMCF} \quad \min{}^* \{(c_1^Tx,...,c_Q^Tx) : x \in \mathbf{X}\} .$$

Theorem 5.6.
If xl is a solution of LMCF then it is also a solution of MMCF.

Proof:

The proposition is shown by assuming the existence of a feasible $x \in \mathbf{X}$ with $Cx < Cx1$ and $Cx \neq Cx1$. Then there must be a component $k \in \{1,\dots,Q\}$ such that $c_k^T x < c_k^T x1$ and $c_j^T x \leq c_j^T x1$ for all $j \neq k$ and $j \in \{1,\dots,Q\}$. This implies $Cx \ll Cx1$ in contradiction to the fact that x1 is a solution of LMCF.

■

The next question is to ask for a method to determine a solution of LMCF. We denote by $C_{[i,j]}$ the column of the matrix C in correspondence to the arc $(i,j) \in A$.

Theorem 5.7.

The following three propositions are equivalent:

(i) x1 is a solution of LMCF,

(ii) in G(x1) exists no cycle L such that
$C(L) = (c_1(L),\dots,c_Q(L)) \ll 0$ where
$c_k(L) := \Sigma_{(i,j)\in L} \, char_{ij}(L) \, c_k(i,j)$ for $k = 1,\dots,Q$;

(iii) there exist price vectors $\pi_i = (\pi_i(1),\dots,\pi_i(Q)) \in \mathbf{R}^Q$;
$i = 1,\dots,n$ such that for all $(i,j) \in A$ it holds
$\pi_i - \pi_j + C_{[i,j]} \gg 0$ implies $x(i,j) = 0$ and
$\pi_i - \pi_j + C_{[i,j]} \ll 0$ implies $x(i,j) = cap(i,j)$.

Proof:

Let x be an optimal solution of LMCF. Assume the existence of a cycle L in G(x) with $C(L) \ll 0$. This implies with $\epsilon > 0$ that for $z := x + \epsilon\cdot char(L)$ it holds $z \in \mathbf{X}$ and $Cz = Cx + \epsilon\cdot C(L) \ll Cx$ in contradiction to the lexicographical optimality of x. On the other side, from the existence of a flow z: $Cz \ll Cx$ we can evaluate the difference flow $dx = z - x$. dx may be represented by means of a set of $m - n + 1$ fundamental cycles $L_1,\dots,L_{m-n+1}$ with $dx = \Sigma_i \, \beta_i\cdot char(L_i)$. Since it is $C\cdot dx \ll 0$, among the fundamental cycles must be at least one cycle L with $C\cdot char(L) \ll 0$. Consequently, no flow z with the above properties may exist. The equivalence between (i) and (iii) is a special case of the general optimality results for lexicographic linear programming (see Isermann 1982).

■

The stated theorem suggests two methods for solving LMCF. In the first one, starting from a feasible flow x, lexicographically negative cycles are determined consecutively. This can be done by any procedure to calculate negative cycles using a modified selection rule only. The second possibility is due to the lexicographic network simplex method. For this sake, in procedure SIMPLEX of Chapter 3 reduced costs are replaced by reduced cost vectors

(14) $Rc(i,j) \;:=\; C_{[i,j]} + \pi_i - \pi_j \in \mathbf{R}^Q$ for all $(i,j) \in A$.

This implies modified sets A_{C+}^* and A_{C-}^* related to the lexicographical ordering:

$$A_{C+}^* := \{(i,j) \in A_{C+}: Rc(i,j) \gg 0\}$$
$$A_{C-}^* := \{(i,j) \in A_{C-}: Rc(i,j) \ll 0\}.$$

5.3.2. Goal Programming

Another approach for solving multicriteria problems is goal programming (Ignizio 1976). Within this approach, the decision maker typically specifies upper and lower levels for the values of each of the objective functions. Subsequently, a single criterion problem with the objective to minimize the weighted derivations of these levels is solved. In the case of the flow polyhedron $\mathbf{X}$ and Q linear objective functions $c_k^T x$; $k = 1,\ldots,Q$ this leads to

$$\text{GP} \quad \min \Big\{ \sum_{k=1}^{Q} (u_k \cdot \mu_k + w_k \cdot \theta_k) : x \in \mathbf{X}$$
$$c_k^T x + \mu_k - \theta_k = b_k \qquad k = 1,\ldots,Q$$
$$\mu_k,\ \theta_k \geq 0 \qquad k = 1,\ldots,Q \Big\}$$

where

- μ_k - negative deviation of objective k
- θ_k - positive deviation of objective k
- u_k, w_k - weights associated with μ_k respectively θ_k
- b_k - aspiration level of objective k.

GP is a network flow problem with Q general linear constraints. Concerning the efficiency of solving this class of problems we refer to Chapter 8.

5.3.3. The Multicriteria Network Simplex Method

The multicriteria simplex method developed by Yu & Zeleny (1975) can be adopted for network flows. In generalization to the single-criterion case we need the vectors $Rc(i,j) \in \mathbf{R}^Q$ of reduced costs (14) to verify the optimality of a given basic solution x which is related to a spanning tree T.

Theorem 5.8.
Let $x \in \mathbf{X}$ be a basic solution with corresponding spanning tree $T = (V,A_T)$. Let π be the $n \times Q$ - matrix of price vectors π_j for all $j \in V$. Then $x \in X_{eff}$ if and only if there is a vector $t \in \mathbf{R}^Q$ such that

(i) $t_k > 0$ for all $k = 1,..,Q$;

(ii) $(i,j) \in A_C$ and $x(i,j) = cap(i,j)$ implies $t^T Rc(i,j) \le 0$;

(iii) $(i,j) \in A_C$ and $x(i,j) = 0$ implies $t^T Rc(i,j) \ge 0$.

Proof:
From Theorem 5.1. we know that $x \in X_{eff}$ if and only if there is a strictly positive vector t such that $x \in \arg\min \{t \cdot Cx : x \in \mathbf{X}\}$. Now we can apply the optimality conditions of the single criterion case and this results in the proposition.

■

Since X_{eff} is known to be connected, we can choose any strictly positive parameter vector $t > 0$ to compute a first efficient basic solution. Specializing Theorem 5.3., we obtain necessary and sufficient conditions that an adjacent basic solution also belongs to X_{eff}. For this sake, the spanning tree $T' = (V,A_T')$ and the matrix π' in correspondence to an adjacent basic solution z are investigated. Furthermore, we use

$A_C' := A - A_T'$,

$A1 := \{(i,j) \in A_C': z(i,j) = 0\}$, and

$A2 := \{(i,j) \in A_C': z(i,j) = cap(i,j)\}$.

Theorem 5.9.
Let $x \in \mathbf{X}$ be an efficient basic solution. For an adjacent basic solution $z \in \mathbf{X}$ it holds $z \in X_{eff}$ if and only if

$$(15)\quad G(e,y) := \max \{1^T e : (e,y) \in \mathcal{L}\} = 0 \quad \text{with } \mathcal{L} \text{ defined by}$$

$$e_k + \Sigma_{(i,j)\in A1} (c^k{}_{[i,j]} + \pi_i'(k) - \pi_j'(k))\, y(i,j) - \Sigma_{(i,j)\in A2} (c^k{}_{[i,j]} + \pi_i'(k) - \pi_j'(k))\, y(i,j) = 0 \quad \text{for } k = 1,...,Q;$$

$$e \ge 0\ ,\ y \ge 0.$$

Proof:
Using $Rc'(i,j)$ for the vector of reduced costs related to z, the dual problem related to (15) can be written as

$$(16)\quad \min \{0^T t: \begin{array}{ll} t^T Rc'(i,j) \ge 0 & \text{for } (i,j) \in A1, \\ t^T Rc'(i,j) \le 0 & \text{for } (i,j) \in A2, \\ t_k \ge 1 & \text{for } k = 1,...,Q\}. \end{array}$$

According to Theorem 5.8., these are the optimality conditions with respect to z. If $G(e,y) > 0$ then there must be at least one component $k \in \{1,...,Q\}$ with $e_k > 0$. Consequently, the optimal value of (15) can be made infinitely large. Due to the duality theory of linear programming , the feasible area of (16) is empty. With $y = 0$ it is shown that the feasible area of (15) is non-empty. If the optimal value is equal to 0 then this implies the existence of a vector t satisfying the conditions of Theorem 5.8., i.e., $z \in X_{eff}$.

■

5.3.4. Interactive Methods

The interaction between the solutions of the model and the preferences of the decision maker is widely used. Kok (1986) defined an interactive method as a formalized information procedure between solutions of a model (in the objective space) and the preferences of a decision maker. With the interactive approach, the decision maker can learn about his preferences, thus obtaining more insight into the conflict situation. Additionally, he can influence the amount of computation, the number of calculated efficient points and also, which solutions are preferred. This needs a permanent exchange of information between decision maker and the model. Mostly, interactive methods are based on scalarizing and a basic routine to solve the optimal solutions of these scalarized problems. At first such a method considers a linear combination of the objective function as described in Theorem 5.1. An initial solution is determined with a strictly positive weighting vector $t \in \mathbf{R}^Q$. This results in a basic solution causing for each $(i,j) \in A_C$ a vector $Rc(i,j) = C_{[i,j]} + \cap_i - \cap_j$ of reduced costs indicating the changes in the objective function when one unit of the non-basic variable $x(i,j)$ enters into the basis. In the method of Zionts & Wallenius (1976), all these trade-off vectors are offered to the decision maker, who may choose which of the tradeoffs are attractive or not.

Another interactive approach is due to (iii) of Theorem 5.1. Now one objective is optimized with all other objective functions in the role of additional linear constraints. In the Interactive Multiple Goal Programming method of Spronk (1981), at each iteration, the ideal and the nadir vector are offered to the decision maker. From there, a conclusion can be made, which objective function should be under a certain threshold level. This leads to an updating of one or more elements of the right side vector.

5.4. An Exact Method for Bicriteria Minimum-Cost Flows

In the case of two goal functions there is the possibility to give a graphical representation of the objective space. For the flow polyhedron **X** defined by the flow conservation rules and the capacity constraints we define $Y := \{(c^Tx, d^Tx): x \in \mathbf{X}\} \subset \mathbf{R}^2$. It is well known that different basic solutions may be mapped into the same extreme point in the objective space. The multicriteria solution methods based on parameter variation and adjacent basic movements are influenced by this fact, since the selection criteria is related to the extreme points in the decision space. Unfortunately, the number of these extreme points may be very large, because there may be extreme points of **X** which does not correspond to an extreme point in the objective space. Additional difficulties may arise due to degeneration which in the case of the flow polyhedron may occur very often (computational results of Bradley, Brown & Graves 1977 indicated that more than 90% of the several thousand pivot steps have been degenerated).

One idea to circumvent these difficulties is to determine extreme points in the objective space. Aneja & Nair (1979) applied this approach to bicriteria transportation problems. We firstly extend their method to the bicriteria minimum-cost flow problem BMCF. In Section 5.5., an approximation method is developed such that the set

(17) $Y_{eff} := \{y: y = (c^Tx, d^Tx), x \in X_{eff}\}$

is bounded by upper and lower approximations in dependence of a given $\epsilon > 0$ representing the accuracy of the approximation.

We develop an exact method called ANNA. Its main idea is illustrated in Figure 5.3. At the beginning, two solutions are determined using lexicographical minimization. During the algorithm, two sets K and L are handled which represent pairs of indices of extreme points in the objective space which may respectively may not have further extreme points between them. For any pair $(r,s) \in L$ we assume without loss of generality that $c^Tx_r < c^Tx_s$. Then the problem

(18) $\min \{\beta \cdot c^Tx + \alpha \cdot d^Tx: x \in \mathbf{X}\}$ with

(19) $\beta := d^T(x_r - x_s)$ and

$\alpha := c^T(x_s - x_r)$

is solved. (18) is the minimum-cost flow problem MCF with a scalar objective function. For each $y \in Y_{eff}$ we obtain one corresponding solution x mapped into y.

As above, we denote by Θ_2(BMCF) the number of efficient extreme points of the problem BMCF. The complexity to solve MCF is abbreviated by T(m,n), where the complexity of one of the algorithms described in Chapter 3 may be replaced instead of T(m,n).

```
procedure ANNA
begin
  compute x_1 ∈ arg lexmin {(c^T x, d^T x): x ∈ X}
  compute x_2 ∈ arg lexmin {(d^T x, c^T x): x ∈ X}
  X_eff := {x_1, x_2}
  Y_eff := {y_1, y_2}  with y_1 := (c^T x_1, d^T x_1), y_2 := (c^T x_2, d^T x_2)
  k := 2
  if y_1 = y_2 then stop
             else begin L := {(1,2)} , E := ∅ end
  while L ≠ ∅ do
  begin
    choose (r,s) ∈ L
    compute α and β due to (19)
    compute x' ∈ arg min {β·c^T x + α·d^T x: x ∈ X}
    y' := (c^T x', d^T x')
    if y' = y_r or y' = y_s then begin
                                   L := L - {(r,s)},
                                   E := E + {(r,s)}
                                   end
                              else begin
                                   k := k + 1,
                                   x_k := x', y_k := y'
                                   X_eff := X_eff + {x_k},
                                   Y_eff := Y_eff + {y_k},
                                   L := L + {(r,k),(k,s)} - {(r,s)}
                                   end
  end
end
```

Theorem 5.10.
Procedure ANNA computes the set Y_{eff} in $O(\Theta_2(\text{BMCF})\cdot T(m,n))$ steps.

Proof:
Due to Theorem 5.8., the membership of y_1, y_2 to Y_{eff} is valid. In the

case that $y_1 = y_2$ the set Y_{eff} has only one element and the procedure terminates. Otherwise, $L = \{(1,2)\}$ and for $(r,s) = \{(1,2)\}$ we obtain that $\beta = y_1(2) - y_2(2) > 0$, $\alpha = y_2(1) - y_1(1) > 0$. Assume that

$$(20) \quad y_k \in \arg\min \{\beta \cdot y(1) + \alpha \cdot y(2) : y = (y(1),y(2)) \in Y\}.$$

Then it is $y_k \in Y_{eff}$ since otherwise with a dominating y' it follows $\beta \cdot y'(1) + \alpha \cdot y'(2) < \beta \cdot y_k(1) + \alpha \cdot y_k(2)$, and this contradicts (20). If y_k is equal to y_r or y_s then because of the convexity of Y we know that y_r and y_s are adjacent extreme points. Otherwise we have to investigate the pairs defined by the indices (r,k),(k,s). Consequently, in each iteration either a new efficient solution is found or the adjacency of two extreme points is demonstrated. The algorithm terminates if $L = \emptyset$. That means, we have calculated all adjacent extreme points in the objective space, and consequently the complete set Y_{eff}.

Each of the subproblems (20) has a complexity $O(T(m,n))$. The procedure needs $2 \cdot \theta_2(BMCF) - 3$ such calls to prove the stated properties. Additionally, two minimum-cost flow problems with lexicographic goal function must be solved. Using

$$w \in \arg \operatorname{lex} \min \{(c^Tx, d^Tx) : x \in \mathbf{X}\} \text{ if and only if}$$
$$w \in \arg\min \{c^Tx + \delta \cdot d^Tx : x \in \mathbf{X}\}$$

for a sufficiently small value $\delta > 0$ results in $2 \cdot \theta_2(BMCF) - 1$ calls of MCF, i.e., the stated complexity is shown.

■

We take the pathological instance of the proof to Theorem 5.5. to illustrate the above procedure. The solution of the lexicographic problems results in

$$x_1 = \sum_{k=1}^{8} z_1(k) \cdot \text{char}(P_k) \quad \text{with } z_1 = (8,0,0,0,0,0,0,0,0) \quad \text{and}$$

$$x_2 = \sum_{k=1}^{8} z_2(k) \cdot \text{char}(P_k) \quad \text{with } z_2 = (0,1,1,1,1,1,1,1,1).$$

We remark that the ordering of the solutions calculated by ANNA does not coincide with the ordering given in Table 5.1. With $L = \{(1,2)\}$ and $y_1 = (8,16)$, $y_2 = (40,8)$ it follows due to (19) that $\alpha = c^T(x_s - x_r) = 8$ and $\beta := d^T(x_r - x_s) = 32$. The solution of min $\{32 \cdot c^Tx + 8 \cdot d^Tx : x \in \mathbf{X}\}$ is x_3 with $z_3 = (4,1,1,1,1,0,0,0,0)$. The objective function vector $y_3 = (18,12)$ is

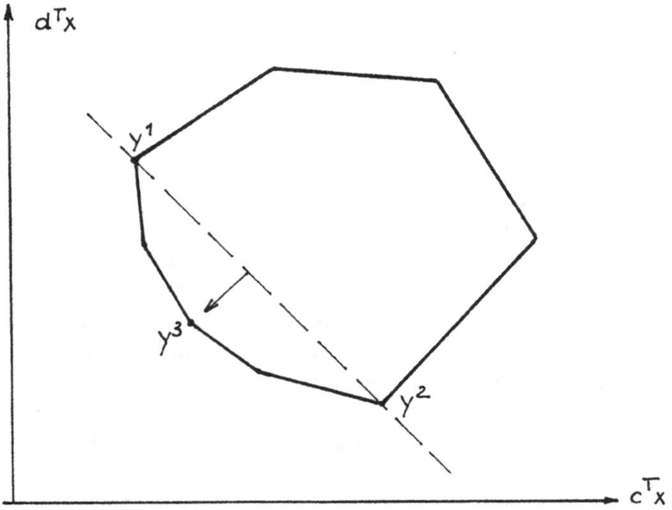

Figure 5.3. Calculation of y_3 due to procedure ANNA.

Table 5.2. Solution of the pathological instance of Theorem 5.5. using ANNA (continuation).

(r,s)	a	b	k	z_k	y_k	L	E
(1,4)	2	3	5	(7,1,0,0,0,0,0,0,0)	(9,15)	(2,3),(3,4) (1,5),(4,5)	∅
(1,5)	1	1				(2,3),(3,4) (4,5)	(1,5)
(4,5)	1	2				(2,3),(3,4)	E := E + (4,5)
(3,4)	2	7	6	(5,1,1,1,0,0,0,0,0)	(14,13)	(2,3),(3,6) (4,6)	
(4,6)	1	3				(2,3),(3,6)	E := E + (4,6)
(3,6)	1	4				(2,3)	E := E + (3,6)
(2,3)	4	22	7	(2,1,1,1,1,1,1,0,0)	(27,10)	(2,7),(3,7)	
(3,7)	2	9	8	(3,1,1,1,1,1,0,0,0)	(22,11)	(2,7),(3,8) (7,8)	
(7,8)	1	5				(2,7),(3,8)	E := E + (7,8)
(3,8)	1	4				(2,7)	E := E + (3,8)
(2,7)	2	13	9	(1,1,1,1,1,1,1,1,0)	(33, 9)	(2,9),(7,9)	
(7,9)	1	6				(2,9)	E := E + (7,9)
(2,9)	1	7				∅	E := E + (2,9)

not equal to y_1 or y_2. So we have found a new extreme point of Y_{eff}. It follows that L := L + {(1,3),(2,3)} - {(1,2)}. For the next iteration, we choose (r,s) = (1,3) and obtain $\beta = 4$, $\alpha = 10$. The solution of the corresponding problem (18) is x_4 with $z_4 = (6,1,1,0,0,0,0,0,0)$. Since y_1, $y_3 \neq y_4$ L is updated to L := L + {(1,4),(3,4)} - {(1,3)}. The subsequent steps of the algorithm are summarized in Table 5.2.

From Theorem 5.5. we know that ANNA may need an exponential running time. It is natural to ask for an approximation of the solution set.

5.5. Approximative Methods for Bicriteria Flows

The set of all efficient points in bicriteria optimization problems may be considered as a convex curve with one of the two objective function values as the independent variable. Burkard, Hamacher & Rote (1987) investigated the approximation of a given convex function g(t) on a bounded interval $[a,b] \subset \mathbf{R}$. They assumed that f is continuous in the endpoints of the interval and that for any $t \in (a,b)$ the left and the right derivatives are available. Additionally, the right derivative in a and the left derivative in b should be finite. They introduced the so-called sandwich algorithm for determining lower and upper piece-wise linear approximations for a convex function with the above given properties. With l(t) and u(t) as lower respectively upper approximation, the inequalities

(21) $l(t) \leq g(t) \leq u(t)$ for all $t \in [a,b]$

must be satisfied.The sandwich algorithm constructs a finite partition

(22) $a = t_1 < t_2 < \ldots < t_h = b$

of the interval [a,b] with the property that the slope of both l(t) and u(t) is constant in each of the subintervals. A list L of those subintervals $[t_i,t_{i+1}]$ of [a,b] is maintained in which the error given by (22) is greater than ϵ. At the beginning, L contains as single element the interval [a,b] or is empty if the error bound is already met. At each iteration, a subinterval $[t_i,t_{i+1}]$ is chosen from the list. A new point t^* with $t^* \in (t_i,t_{i+1})$ is constructed combined with an updating of l(t) and u(t). Now it is checked whether in the two new intervals $[t_i,t^*]$ and $[t^*,t_{i+1}]$ the error bound is met. If not, the corresponding interval(s) are added to the list. The sandwich algorithm terminates with ϵ-approximative functions $l(t)$ and $u(t)$ if L is empty. If t^* is taken as the midpoint of the interval $[t_i,t_{i+1}]$ (bisection rule), then at each iteration the solution of a minimum-cost flow problem with one additional linear constraint is necessary. Originally, the maximum error rule

(23) $\max \{u(t) - l(t): t \in [a,b]\}$

was taken as a measure of the quality of the approximation. However, this measure is not invariant under rotations of the functions. An alternative measure was considered in Ruhe (1988a). Using

$$R := \{(t,l(t)): t \in [a,b]\},\ S := \{(t,u(t)): t \in [a,b]\},$$

and $\delta(x,y)$ for the euclidean distance between two points x and y, the *projective distance* pdist(l(t),u(t),[a,b]) between l(t), u(t) in the interval [a,b] is defined as

(24) $\text{pdist}\ (l(t),u(t),[a,b]) := \max\ \{pd1,pd2\}$ with

$$pd1 := \sup_{x\in R} \inf_{y\in S} \delta(x,y)$$

$$pd2 := \sup_{y\in S} \inf_{x\in R} \delta(x,y).$$

As a second modification, another rule for partitioning the intervals which do not meet the error bound was introduced. According to this rule, the partitioning is determined by that point of the function g(t) where the supporting line is parallel to the line connecting the two endpoints of the curve in the considered interval. This will be called *chord-rule*. In the following, a more detailed description of the approximation algorithm using projective distance and chord-rule when applied to bicriteria minimum-cost flows is given.

Analogously to the exact algorithm, the two solutions x_1,x_2 resulting from lexicographical minimization due to

(25) $x_1 \in \text{lex min}\ \{(c^Tx,d^Tx): x \in \mathbf{X}\}$

(26) $x_2 \in \text{lex min}\ \{(d^Tx,c^Tx): x \in \mathbf{X}\}$

are determined. To constitute first upper and lower approximations, the right respectively left derivatives of the function g(t) defined by the set of points $\{(c^Tx,d^Tx): x \in X_{eff}\}$ is necessary. With the values of the first objective function as independent variable, the parameter interval is $[a,b] = [c^Tx_1,c^Tx_2]$. We assume that [a,b] is partitioned as in (22). This corresponds to a partition

$$0 < \Gamma_1 < \ldots < \Gamma_h < \infty$$

of the parameter interval solving

$$(27) \quad \min \{c_1^T x + \Gamma \cdot c_2^T x : x \in \mathbf{X}\} \qquad \text{for } \Gamma \in (0,\infty).$$

Lemma 5.4.

Let x_1, x_i; i= 2,...,h, and x_{h+1} be optimal solutions of (27) for the intervals $(0,\Gamma_1]$, $[\Gamma_{i-1},\Gamma_i]$; i = 2,...,h and $[\Gamma_h,\infty)$, respectively. Then it holds

(i) $g^+(a) = -1/\Gamma_1$

(ii) $g^-(t_i) = -1/\Gamma_{i-1}$ for i = 2,...,h

(iii) $g^+(t_i) = -1/\Gamma_i$ for i = 2,...,h , and

(iv) $g^-(b) = -1/\Gamma_h$.

Proof:

It is well known that each basic solution $x \in X_{eff}$ implies a nonempty polyhedral region $\Omega(x)$ containing all parameter vectors $w \in \mathbf{R}^Q$ such that $x \in \arg\min \{w \cdot Cx: x \in \mathbf{X}\}$. The vectors $w_1 = (1,\Gamma_1)$ and $w_2 = (1,\Gamma_h)$ define upper respectively lower boundary constraints of $\Omega(x_1)$ and $\Omega(x_h)$. The derivations $g^+(a)$ and $g^-(b)$ are given by the slope of the corresponding constraints, i.e., it holds (i) and (iv). The vectors $(1,\Gamma_{i-1})$ and $(1,\Gamma_i)$ corresponding to $x_i \in X_{eff}$ are the upper and lower boundary constraints for $\Omega(x_i)$; i = 2,...,h. By the same arguments as above we obtain propositions (ii) and (iii).

■

The upper and lower approximation of g(t) in the interval [a,b] is initialized according to

$$(28) \quad u^1(t) := g(a) + [(g(b)-g(a))/(b-a)] \cdot (t-a) \quad \text{and}$$

$$(29) \quad l^1(t) := \max \{g(a) + g^+(a) \cdot (t-a), g(b) + g^-(b) \cdot (t-b)\}.$$

Taking $t^* \in \arg\min \{[f(a)-f(b)] \cdot t + (b-a) \cdot g(t): t \in [a,b]\}$ we produce improved approximations

$$(30) \quad u^2(t) := \begin{cases} g(a) + [(g(a)-g(t^*))/(a-t^*)](t-a) & \text{for } t \in [a,t^*] \\ g(b) + [(g(t^*)-g(b))/(t^*-b)](t-b) & \text{for } t \in [t^*,b] \end{cases}$$

and

$$(31) \quad l^2(t) := \begin{cases} g(a) + g^+(a)(t-a) & \text{for } t \in [a,\underline{a}] \\ g(\underline{a}) + [(g(a)-g(b))/(a-b)](t-\underline{a}) & \text{for } t \in [\underline{a},\underline{b}] \\ g(b) + g^-(b)(t-b) & \text{for } t \in [\underline{b},b] \end{cases}$$

where $\underline{a},\underline{b}$ are determined by the cutting points of $l^1(t)$ with $h(t) := g(t^*) + [(g(a)-g(b))/(t^*-b)](t-t^*)$.

The construction of the first and second approximation of u(t) and l(t) are illustrated in Figure 5.4.

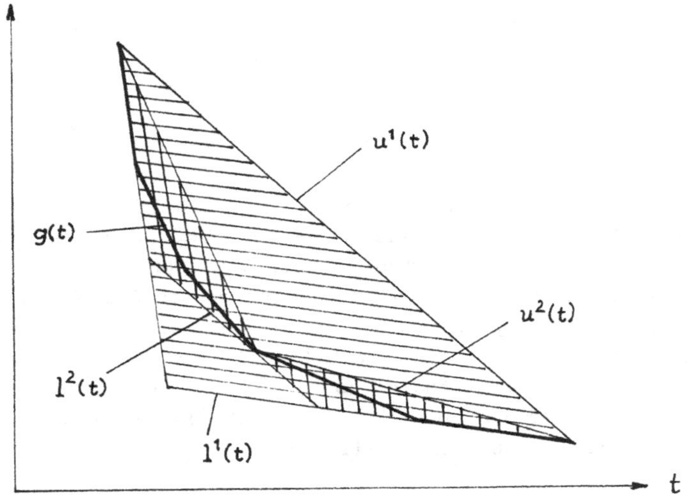

Figure 5.4. Approximations $u^1(t), l^1(t)$ and $u^2(t), l^2(t)$ of $g(t)$.

In the following lemma, we consider a parameter interval $[t_1,t_2]$ such that the corresponding triangle defined by l(t) and u(t) with $u(t_1) = l(t_1)$, $u(t_2) = l(t_2)$ and one breakpoint in l(t) does not meet the error bound. The question is for the degree of improvement which is caused by one iteration.

Lemma 5.5.
Let $l^1(t)$, $l^2(t)$ and $u^1(t)$, $u^2(t)$ be approximating functions of g(t) in an interval $[t_1,t_2]$ as defined in (28) - (31). Then

$$\tau^2 := \text{pdist}(l^2(t),u^2(t),[t_1,t_2]) < \text{pdist}(l^1(t),u^1(t),[t_1,t_2])/2 =: \tau^1/2.$$

Proof:

We use the notation as introduced in Figure 5.5. and firstly proof that $\delta(E,G) \le \delta(C,D)/2$. Assume J located at the line l(C,D) such that $\delta(C,J) = \delta(J,D)$. If $\delta(C,I) \ge \delta(C,J)$ then $\delta(E,G) < \delta(I,D) \le \delta(C,D)/2$. Similarly, $\delta(C,K) \le \delta(C,J)$ then $\delta(E,G) < \delta(C,K) \le \delta(C,D)/2$. Now let be $\delta(C,I) < \delta(C,J) < \delta(C,K)$. Consider the line l(A,J) with crossing point P on l(E,G) and $\delta(E,P) = \delta(P,Q)$. Then

$$\begin{aligned}\delta(E,G) &= \delta(E,P) + \delta(P,G) = \delta(P,Q) + \delta(P,G)\\ &= \delta(J,R) + \delta(P,G) < \delta(J,R) + \delta(J,K) < \delta(J,D) = \delta(C,D)/2.\end{aligned}$$

The remaining inequalities are proved analogously.

■

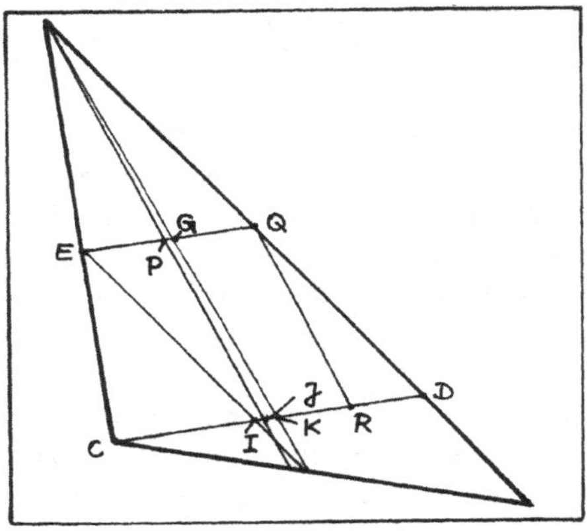

Figure 5.5. Illustration to the proof of Lemma 5.5.

```
procedure APPROX [a,b]
begin
  u(t) := g(a) + [(g(b)-g(a))/(b-a)](t-a)
  l(t) := max {g(a) + g⁺(a)(t-a),g(b) + g⁺(b)(t-b)}
  τ := pdist(l(t),u(t),[a,b])
  if D ≤ ε then stop
           else
           begin
             α := b - a
             β := g(a) - g(b)
             x* := arg min {β·cᵀx + α·dᵀx: x ∈ X}
             t* := cᵀx*
             APPROX [a,t*]
             APPROX [t*,b]
           end
end
```

Let τ^1 := pdist($l^1(t)$,$u^1(t)$,[a,b]) be the value of the initial projective distance between $l^1(t)$ and $u^1(t)$.

Theorem 5.11.
Procedure APPROX determines in $O((\tau^1/\epsilon)T(m,n))$ steps two functions l(t), u(t) such that for a given $\epsilon > 0$ and interval [a,b] the optimal value function $g(t) = \{(c^Tx,d^Tx): x \in X_{eff}\}$ is approximated with

(i) pdist(l(t),u(t),[a,b]) $\le \epsilon$ and

(ii) $l(t) \le g(t) \le u(t)$ for all $t \in [a,b]$.

Proof:
APPROX terminates if $\tau \le \epsilon$. Consider index $p:=\min\{k \in \mathbf{N}: \tau^1 \cdot 2^{-k} \le \epsilon\}$. At each iteration we have to solve a problem MCF and must calculate the cutting point of two lines. To achieve $\tau \le \epsilon$, this needs at most $Q = 2^p - 1$ iterations. The complexity T(m,n) to solve MCF is dominating in each iteration. Using $Q = O(\tau^1/\epsilon)$ results in the complexity bound for the whole procedure. (ii) is fulfilled for $u^1(t)$, $l^1(t)$ in [a,b]. From the recursive character of the procedure and the rules for updating the approximations, this property is transferred to all subsequent intervals.

■

The procedure APPROX will be illustrated by the pathological example of Theorem 5.5. for G_3 and $\epsilon = 1$. At the beginning, we take the solutions x_1 and x_2 obtained from lexicographic minimization (25),(26). So we have two points in the objective space $y_1 = (8,16)$, $y_2 = (40,8)$ and an interval $[a,b] = [8,40]$. The function defined by y_1 and y_2 is $u^1(t) = -t/4 + 18$. For the construction of $l^1(t)$ we need $g^+(8)$ and $g^-(40)$. From Table 5.1. we know that x_1 is an optimal solution of the parametric problem (13) for $\mu \in (0,1/2]$. It follows that $g^+(8) = -1/\Gamma_1 = -1$. Analogously, we compute $g^-(40) = -1/7$. The cutting point of both supporting lines is for $t = 12$. Consequently,

$$l^1(t) = \begin{cases} 24 - t & \text{for } t \in [8,12] \\ (96-t)/7 & \text{for } t \in [12,40]. \end{cases}$$

As a measure of the quality of the approximation we calculate pdist($l^1(t)$,$u^1(t)$,[8,40]) = $12/\sqrt{17}$. From the solution of $\min\{32 \cdot c^Tx + 8 \cdot d^Tx : x \in \mathbf{X}\}$, a new point $y_3 = (18,12)$ is obtained. This leads to improved approximations

$$u^2(t) = \begin{cases} -2t/5 + 96/5 & \text{for } t \in [8,18] \\ -2t/11 + 168/11 & \text{for } t \in [18,40]. \end{cases}$$

$$l^2(t) = \begin{cases} 24 - t & \text{for } t \in [8,10] \\ -t/4 + 33/2 & \text{for } t \in [10,26] \\ (96-t)/7 & \text{for } t \in [26,40]. \end{cases}$$

At this stage of the approximation it is pdist($l^2(t)$,$u^2(t)$,[8,18]) = 1 and pdist($l^2(t)$,$u^2(t)$,[18,40]) = $6\cdot\sqrt{5}/25 < 1$. Since the maximum of both distance values is not greater than 1, the procedure terminates. The result is a set of three efficient solutions representing the whole set in the sense that all remaining efficient points in the objective space are contained in the area included by the upper and lower approximation.

Theorem 5.12. (Rote 1988)
Suppose an interval [a,b] of length L = b - a, where the function values and the one-sided derivatives $g^+(a)$ and $g^-(b)$ have been evaluated. Let the slope difference be $\sigma := g^-(b) - g^+(a)$. Then, in order to make the greatest vertical error between the upper and the lower approximation smaller than or equal to ϵ, APPROX combined with chord-rule needs at most $m_\epsilon(L\cdot\sigma)$ additional evaluations of $g(x)$, $g^+(x)$, and $g^-(x)$, where

$$m_\epsilon(L\cdot\sigma) := \begin{cases} 0 & \text{for } L\cdot\sigma \le 4, \\ \lfloor\sqrt{(L\cdot\sigma/\epsilon)} - 2\rfloor & \text{for } L\cdot\sigma > 4. \end{cases}$$

■

Fruhwirth, Burkard & Rote (1989) proposed the angle bisection rule and the slope bisection rule instead of the chord rule. In these rules, the new direction is chosen either as the angle bisector of the two known supporting lines at the endpoints, or as the direction the slope of which is the midpoint of the slopes of those two lines. Both rules are illustrated in Figure 5.6.

Theorem 5.13. (Fruhwirth, Burkard & Rote 1989)
Assume L_0 to be the length of $l(t)$ at the initialization of APPROX. The number M of efficient points needed to obtain an ϵ-approximation of the optimal value function g(t) is bounded by

(i) $M \le \max\{\lfloor 3/2\sqrt{(\pi\cdot L_0/4\epsilon)}\rfloor + 1, 2\}$ for angle bisection and
(ii) $M \le \max\{\lfloor 3/2\sqrt{(L_0/\epsilon)}\rfloor + 1, 2\}$ for slope bisection.

■

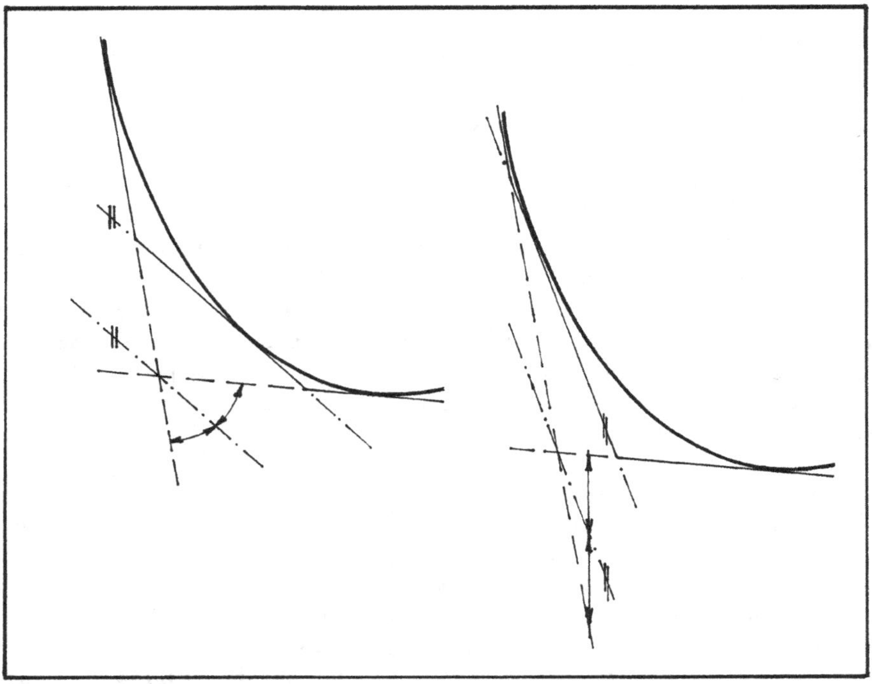

Figure 5.6. (a) Illustration of angle bisection rule.
(b) Illustration of slope bisection rule.

5.6. ε-Optimality

One of the crucial points in multi-criteria decision making is bounding the number of (efficient) solutions. One of the possible answers to this situation is to approximate the set of Pareto-optimal solutions. In this part, a new approximation approach based on the notion of ε-optimality is described. Instead of focusing on the Pareto-optimality of specific feasible solutions, we consider the range of the two objective functions. A pseudopolynomial algorithm using a lower and an upper approximation of the curve of all efficient points in the objective space is developed.

Searching for ϵ-optimal instead of optimal solution sets has three advantages:

(i) It is possible to restrict the number of solutions from the very beginning while the calculated solution sets represent the whole set in a reasonable way.

(ii) An accuracy ϵ can be incorporated such that for any feasible solution there is one from the optimal solution set with the property that the maximum relative improvement is bounded by $\epsilon/(1+\epsilon)$.

(iii) The time to calculate ϵ-optimal solutions sets is much shorter compared to the computation of the whole set.

Let $x = (x_1,x_2,\ldots,x_K)$, $y = (y_1,y_2,\ldots,y_K)$ be positive K-vectors and suppose an accuracy $\epsilon \geq 0$. Then x *ϵ-dominates* y if

$x_k \leq (1 + \epsilon)\cdot y_k$ for all $k = 1,\ldots,K$.

A subset $S \subset \mathcal{X}$ of feasible solutions is called *ϵ-optimal* with respect to a vector valued function $f: \mathcal{X} \longmapsto Y \subset R^K$ if for every feasible solution $x \in \mathcal{X} - S$ there is a solution z_x such that $f(z_x)$ ϵ-dominates $f(x)$, i.e.,

(32) $f_k(z_x) \leq (1+\epsilon)\cdot f_k(x)$ for all $k = 1,\ldots,K$.

In the case of linear programming as considered here, we associate with a given sequence $X^0 = \{x_1,\ldots,x_p\}$ of feasible solutions a set $S(X^0)$ defined due to

$S(X^0) := \{x \in \mathcal{X}: x = t\cdot x_{i-1} + (1-t)\cdot x_i,\ 0 \leq t \leq 1,\ i = 2,\ldots,p\}$.

In close relationship to approximations calculated by APPROX, two questions arise:

(i) Assume the cardinality card(S) = s of S is given and APPROX is applied with the supplement that at each iteration the interval of maximum error is partitioned. Which is the minimum ϵ-value such that the calculated set has cardinality s and is ϵ-optimal?

(ii) Assume that $\epsilon > 0$ and the cardinality card(S) = s are given. Compute an ϵ-optimal set of small cardinality using the lower and upper approximations of algorithm APPROX.

To answer the first question, we assume the existence of l(t) and u(t) as defined in connection with APPROX. Let $X_\epsilon = \{x_1,\ldots,x_s\}$ be the set of efficient solutions obtained from the application of the above

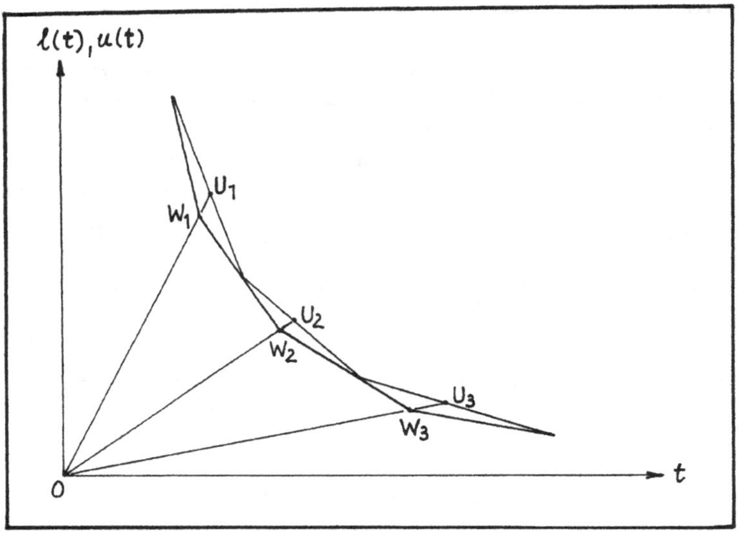

Figure 5.7. Construction of U_i from the inner breakpoints W_i.

procedure. Assume that the solutions of X_ϵ are ordered such that

$$c^T x_1 < c^T x_2 < \ldots < c^T x_s.$$

Consider the objective space: the set $S(X_\epsilon)$ is mapped into the upper approximating function $u(t)$. Let be $W_1,\ldots,W_{s-1}$ be the inner breakpoints of $l(t)$. Each of these points when connected with the origin O defines a line having crossing points $U_1,\ldots,U_{s-1}$ with $u(t)$. This is illustrated in Figure 5.7.

Taking $z = (t,u(t))$ and $\delta(z) := [(t^2 + u^2(t))]^{1/2}$ for the euclidean distance of z from the origin, we can formulate:

Theorem 5.14.
With $X_\epsilon = \{x_1,\ldots,x_s\}$, $l(t)$, $u(t)$ as generated by the application of APPROX and points $U_1,\ldots,U_{s-1}$ on $u(t)$ derived from the inner breakpoints of $l(t)$ the set $S(X_\epsilon)$ is ϵ-optimal for

$$\epsilon = \max \{\delta(U_i,W_i)/\delta(O,W_i): 1 \le i \le s - 1\}.$$

Proof:

For all $x \in \mathcal{L}$ whose image is not located below u(t) there is a solution $z_x \in S(X_\epsilon)$ with $f(z_x) \leq f(x)$. The lower approximation function l(t) ensures that there can be no element x^* with $f(x^*)$ below l(t). Consider a solution $x \in \mathcal{L}$ the image $f(x) = (c^Tx, d^Tx)$ of which is located in the area between u(t) and l(t). The line connecting a point f(x) with the origin has crossing points U_x and W_x with u(t) and l(t), respectively. From geometrical reasons it follows that for maximizing the ratio $\delta(U_x,W_x)/\delta(O,W_x)$ it is sufficient to consider the breakpoints of l(t) and the corresponding crossing points on u(t). With z_x such that $f(z_x) = U_x$ constructed from f(x) it holds $f(z_x) \leq (1+\epsilon)f(x)$. ∎

Now we investigate the question stated in (ii). We define a procedure called ϵ-SANDWICH characterized by the solution of two subproblems per iteration. Typically, at each iteration a triangle violating the given accuracy is investigated, and we ask for the improvement which is caused by one iteration. We give a rule how to compute two further efficient solutions such that the error of each of the three resulting triangles is at most half the error of the considered one.

With an interval [a,b] under consideration the triangle is characterized by the vertices (a,l(a)), (b,l(b)) with l(a) = u(a), l(b) = u(b), and the breakpoint W of l(t). The definition of efficiency implies that the angle at W is greater than $\pi/2$. The line from the origin through W crosses u(t) in U. On this line, there is a point V = (V(1),V(2)) such that $\delta(W,V) = \delta(V,U)$. The two gradients are:

(33) $h^1 := (V(2) - l(a), a - V(1))$

(34) $h^2 := (l(b) - V(2), V(1) - a)$.

The construction of h^1 and h^2 is depicted in Figure 5.8.

We give a formal description of ϵ-SANDWICH starting with lower and upper approximations $l^1(t)$, $u^1(t)$ over the interval [a,b] as stated in (28),(29). We use $Y^1 = (a,g(a))$, $Y^2 = (b,g(b))$ and denote by l(A,B) the line defined by the two points A,B. Assume that W is the breakpoint of l(t) in [a,b] and U is the C crossing point of U(t) with the line l(O,W).

```
procedure ε-SANDWICH (l(t),u(t),[a,b])
begin
  if δ(W,U)/δ(O,W) > ε then
  begin
    determine the gradients h^1,h^2 due to (33),(34)
    compute y^p ε arg min {(h^1)^T y: y ε Y}
    compute y^q ε arg min {(h^2)^T y: y ε Y}
    Y_ε := Y_ε + {y^p,y^q}
    m_1 := (y^1(2)-V(2))/(y^1(1)-V(1))
    l(t) := max {l(t),y^p(2) + m_1(t-y^p(1))} in [y^1(1),y^p(1)]
    u(t) := y^1(2) + [(y^p(2)-y^1(2))/(y^p(1)-y^1(1))](t-y^1(1))
                                                in [y^1(1),y^p(1)]
    ε-SANDWICH (l(t),u(t),[y^1(1),y^p(1))
    m_2 := (y^2(2)-V(2))/(y^2(1)-V(1))
    l(t) := max {y^p(2) + m_1(t-y^p(1),y^q(2) + m_2(t-y^q(1)}
                                                in [y^p(1),y^q(1)]
    u(t) := y^q(2) + [(y^p(2)-y^q(2))/(y^p(1)-y^q(1))](t-y^q(1))
                                                in [y^p(1),y^q(1)]
    ε-SANDWICH (l(t),u(t),[y^p(1),y^q(1)])
    l(t) := max {l(t),y^q(2) + m_2(t-y^q(1))} in [y^q(1),y^2(1)]
    u(t) := y^2(2) + [(y^q(2)-y^2(2))/(y^q(1)-y^2(1))](t-y^2(1))
                                                in [y^q(1),y^2(1)]
    ε-SANDWICH (l(t),u(t),[y^q(1),y^2(1)])
  end
end
```

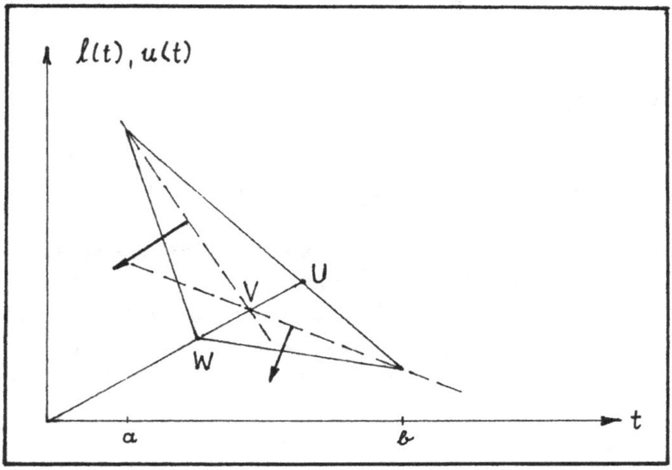

Figure 5.8. The two supporting lines in ε-SANDWICH.

Lemma 5.6. (Menelaos, see Blaschke 1954, p.24)
Assume a triangle A,B,C and a line g. Let P^1,P^2, and P^3 be the crossing points of g with the lines l(A,B), l(B,C), and l(C,A), respectively. Under the assumption that $P^i \neq A,B,C$ for i = 1,2,3 it holds

$$[\delta(A,P^1)/\delta(P^1,B)]\cdot[\delta(B,P^2)/\delta(P^2,C)]\cdot[\delta(C,P^3)/\delta(P^3,A)] = 1.$$

■

Lemma 5.7.
Assume approximating functions $u^1(t)$ and $l^1(t)$ forming a triangle over an interval [a,b], and let ϵ^1 be the minimal value such that the solution set whose image is $u^1(t)$ is ϵ-optimal. One iteration of ϵ-SANDWICH results in improved approximations $u^2(t)$, $l^2(t)$ such that the solution set whose image is $u^2(t)$ is ϵ^2-optimal with $\epsilon^2 < \epsilon^1/2$.

Proof:
We consider the situation shown in Figure 5.9. Firstly, we observe that $\epsilon^1 = \max\ \{\delta(U_i,W_i)/\delta(O,W_i): i = 1,2,3\}$. We investigate the triangle defined by the points O, and U_1 and consider the ratios caused by the crossing points Z_1, W_1, respectively W of the line $l(W,Y_1)$ with the three defining lines of the triangle. From the application of Lemma 5.6. we obtain

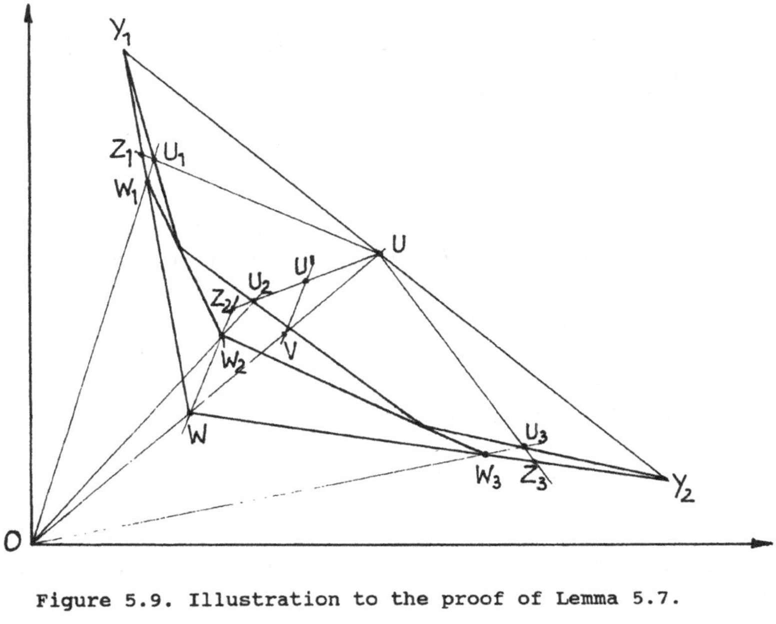

Figure 5.9. Illustration to the proof of Lemma 5.7.

$$\delta(U,W)/\delta(O,W) = \epsilon^1 = \alpha_1 \cdot \delta(U_1,W_1)/\delta(O,W_1) \quad \text{with}$$
$$\alpha_1 = \delta(U,Z_1)/\delta(U_1,Z_1).$$

Since U_1 is below the line $l(Y_1,V)$ (compare (33)!) it follows that $\alpha_1 > 2$. With the same arguments we can prove that

$$\delta(U,W)/\delta(O,W) = \epsilon^1 = \alpha_3 \cdot \delta(U_3,W_3)/\delta(O,W_3) \quad \text{with}$$
$$\alpha_3 = \delta(U,Z_3)/\delta(U_3,Z_3) > 2.$$

It remains to show that the corresponding relation is valid also for the inner triangle. Assume w.l.o.g. that Z_2 is located on the left side related to $l(O,U)$. Then it follows that the slope $m(U,U_2)$ of the line $l(U,U_2)$ is greater than $m(V,W_2)$. $\delta(U,V) = \delta(V,W)$ implies $m(V,U_2) < m(W,Z_2)$. The parallel movement of the line $l(W,Z_2)$ through V results in a crossing point U' on $l(U,U_2)$ with $\delta(U,U') < \delta(U,U_2)$. Because of $\delta(U,U') = \delta(U',Z_2)$ it follows that $\alpha_2 = \delta(U,Z_2)/\delta(U_2,Z_2 > 2$.
Consequently, $\epsilon^2 = \max\{\delta(U_i,W_i)/\delta(O,W_i):\ i = 1,2,3\}$
$= \max\{\epsilon^1/\alpha_i : i = 1,2,3\} < \epsilon^1/2$.

■

Using the above lemma, we can prove a pseudopolynomial behaviour of the approximation approach. We assume that first upper and lower approximations $u^1(t)$ and $l^1(t)$ of $g(t) = \{(c^Tx,d^Tx): x \in X_{eff}\}$ are given as formulated in (28) and (29), respectively. Assuming W to be the breakpoint of $l^1(t)$ and U to be the crossing point of $u^1(t)$ with $l(O,W)$ and taking

(35) $\epsilon^1 = \delta(U,W)/\delta(O,W)$ we can formulate

Theorem 5.15.
For a given $\epsilon > 0$ we assume that procedure ϵ-SANDWICH computes iteratively upper and lower approximations $u(t)$ respectively $l(t)$ of $g(t)$. Then an ϵ-optimal set $S(X_\epsilon)$ is determined in $O((\epsilon^1/\epsilon)^{\log_2 3} \cdot T(m,n))$ steps, where ϵ^1 is given as in (35).

Proof:
Let k be a minimum integer such that $\epsilon^1/2^k \le \epsilon$. Without loss of generality we take

$$2^k = \epsilon^1/\epsilon.$$

For k = 0 we have to solve two lexicographical minimum-cost flow problems resulting in y^1 and y^2. For k = 1, additionally, two problems

MCF (and some computations dominated by the complexity T(m,n)) are necessary. From Lemma 5.7 we know that the error bound is halved at each triangle under consideration. The number of subproblems which must be solved to halve the overall error is at most three times the number of triangles of the previous iteration. This process is repeated iteratively resulting in maximal $3^k - 1$ additional calls of MCF. Taking $k = \log_3(\epsilon^1/\epsilon)\cdot\log_2 3$ leads to the worst case bound stated in the proposition. With the set $S(X_\epsilon)$ of the k-th iteration of ϵ-SANDWICH it holds ϵ-optimality due to Theorem 5.14.

■

5.7. Computational Results

Based on computer implementations of procedures ANNA, APPROX, and ϵ-SANDWICH, some numerical results are given in the following. This concerns the cardinality of Y_{eff}, a comparison of different partition rules in the approximation algorithm, cpu-times , and the performance of the ϵ-optimality approach.

For the cardinality of Y_{eff} we firstly investigated the Zadeh-graphs described in 5.2. with randomly generated costs. The cost coefficients are taken from intervals [0,5], [0,10], and [0,50]. While from Theorem 5.5. an exponential worst-case behaviour for costs as defined in (9) and (10) is known, the number of efficient extreme points for randomly generated costs as presented in Table 5.3. is rather low.

Table 5.3. Average cardinality of Y_{eff} for the Zadeh-graphs G_n with randomly generated costs (five examples in each case).

n	#(V)	#(A)	[0,5]	[0,10]	0,50]
3	8	14	1.6	2.0	1.8
4	10	22	2.2	2.4	1.8
5	12	32	2.2	2.6	1.4

Similar investigations have been taken for grid-graphs with capacities as shown in Figure 5.10.

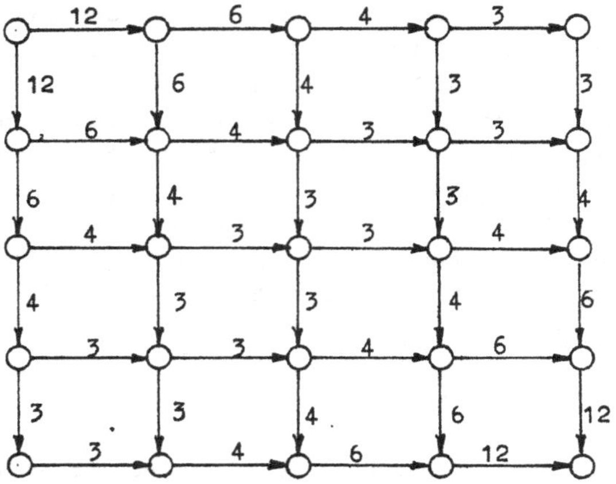

Figure 5.10. Grid-graph. The number on each arc indicates its capacity.

Grid-graphs allow an arrangement of the vertices according to the entries of a matrix such that arcs of the graphs are defined for adjacent (horizontal or vertical) matrix elements only. The capacities of the arcs are defined in such a way that the cuts along the diagonals all have the same capacity value. The costs were generated randomly from different intervals again.

Table 5.4. Cardinality of Y_{eff} for grid-graphs with randomly generated costs.

#(V)	#(A)	[0,5]	[0,10]	[0,20]
16	24	5.6	4.0	7.0
25	40	8.4	10.8	9.6

In Table 5.5., the parameters of the randomly generated test examples using NETGEN (see Klingman, Napier & Stutz 1974) are summarized. Additionally, for varying the arc costs in the interval [1,500] and the capacities in [10,500] , the cardinality θ_2(BMCF) of the set Y_{eff} is given.

Table 5.5. Parameters of the investigated test problems (#(x) - value

Table 5.5. Parameters of the investigated test problems (#(x) - value of the circulating flows) and number θ_2(BMCF) of efficient extreme points.

#	#(V)	#(A)	#(x)	θ_2(BMCF)
1	100	900	600	143
2	100	1000	900	189
3	100	1000	4000	280
4	400	4000	3600	853
5	400	4000	3600	863
6	400	4000	15000	1286
7	800	8000	6000	1539
8	800	8000	12500	2179
9	800	8000	30000	2872

In order to compare different partition rules of the Sandwich algorithm, Fruhwirth, Burkard & Rote (1989) made some numerical investigations using randomly generated examples of different size. They investigated both the angle bisection and the chord rule. Additionally, a strategy combining a left-to-right approach with the angle bisection rule was tested. As a subroutine for solving the minimum-cost flow problem, the primal network simplex code of Ahrens & Finke (1980) was used. In general, primal codes are well suited in this context, because there is to solve a sequence of consecutive minimum-cost flow problems differing in the costs only. So the actual optimal solution may be used as starting solution for the next step. The binary tree corresponding to the interval partitions was treated in a depth-first-search way. The list of intervals that do not meet the error bound is organized as a stack S. For every interval on stack S the efficient solution corresponding to the left endpoint of the interval is stored for later treatment. The index of the up-most topelement of S during the execution of the algorithm is taken as a measure of the memory requirements storing basic solutions. Especially in respect to this measure, the already mentioned left-to-right approach (algorithm LTR) has shown to be very suited.

The algorithm LTR starts by computing x_1, x_2 and an upper bound on the number M (see Theorem 5.15.) of iterations for the angle bisection rule. From the computational results it is known that the bound is not sharp. As a consequence, the authors assumed that the average angle between adjacent line segments of the lower approximative function is given by $F \cdot \alpha_0/M$, where the factor $F > 1$ is determined empirically. Now the objective function is rotated counterclockwise by $F \cdot \alpha_0/M$ using x_1 as starting solution. This leads to an efficient solution x_3 with objective function vector $(c^T x_3, d^T x_3)$ quite close to $(c^T x_1, d^T x_1)$. In general, the error in the interval $[c^T x_1, c^T x_3]$ should be less than ϵ, and the error in the interval $[c^T x_3, c^T x_2]$ will still be quite high. The authors proceed to compute the angle α_3 between the two line segments of the current lower approximative function in $[c^T x_3, c^T x_2]$ and the bound on the number M_3 of iterations necessary to approximate the efficient point function if angle bisection rule would be used. Again, the area Y of all points in the objective space is rotated counterclockwise by $F \cdot \alpha_3/M_3$. Starting from the left line segment of the lower approximative function in the interval $[c^T x_3, c^T x_2]$ leads to another efficient solution. This process is repeated until the rightmost efficient extreme point is reached.

The programs were written in Fortran and the computational experiments were made on a VAX 11/785 at the Graz University of Technology.

For approximating the efficient point function of the bicriteria minimum-cost flow problem, from the results of Fruhwirth et al. (1989) it follows that the LTR method seems to be superior when compared with angle bisection or chord rule. The main reason is that the LTR method solves a sequence of problems which are closely related in the sense that the optimal solution of an actual subproblem is a good starting solution to the next subproblem. Under the assumption $m = 10n$, the LTR method needs about 70% of data memory of the angle bisection and chord rule variant. The experiments also indicated that the number of breakpoints of the efficient point function is far from being exponential.

In Table 5.6., angle bisection, chord rule and LTR-strategy are compared when used in APPROX for the problems characterized in Table 5.5. Each method was tested with different values for ϵ. The given results are due to Fruhwirth, Burkard & Rote (1989). The cpu-times are seconds on VAX 11/785.

Table 5.6. efp - number of efficient points used for ϵ-approximation
sth - height of the upmost element of the stack
$pdist^1 = pdist(u^1(t), l^1(t), [a,b])$

		Angle bisection			Chord rule			LTR-strategy		
#	$\epsilon/pdist^1$	efp	cpu	sth	efp	cpu	sth	efp	cpu	sth
1	.01	13	3.8	3	15	4.0	3	13	3.2	2
	.001	37	6.9	4	33	6.8	5	40	6.0	2
	.0001	85	12.8	5	87	13.3	6	88	11.3	2
2	.01	14	5.0	3	16	5.0	3	13	3.6	2
	.001	38	8.1	5	35	7.6	5	37	6.2	2
	.0001	96	15.2	5	98	15.7	6	98	12.7	2
3	.01	15	5.2	3	16	5.2	3	14	4.2	2
	.001	39	8.6	5	38	8.3	5	40	6.9	2
	.0001	101	16.2	6	100	15.7	6	105	13.1	2
4	.01	13	31.7	3	16	36.3	3	14	26.4	2
	.001	43	57.6	5	39	55.3	5	48	44.0	2
	.0001	124	105.9	6	121	103.6	6	131	79.4	2
5	.01	14	34.4	3	16	36.2	3	13	25.7	2
	.001	42	58.8	5	39	54.5	5	46	43.5	2
	.0001	123	108.3	6	123	107.9	6	130	83.5	2
6	.01	16	41.3	3	17	44.7	4	13	30.7	2
	.001	43	65.8	5	35	61.4	5	42	45.7	2
	.0001	129	118.1	4	127	120.8	6	126	84.3	2
7	.01	13	91.9	3	16	102.7	3	14	74.6	2
	.001	43	156.4	5	38	146.9	5	47	112.4	2
	.0001	123	262.6	6	125	277.3	7	139	202.2	2
8	.01	13	106.0	3	17	123.8	4	15	93.0	2
	.001	43	181.8	5	36	166.7	5	44	127.1	2
	.0001	130	310.1	7	129	316.7	7	146	224.5	2
9	.01	14	136.0	3	16	146.2	3	13	104.5	2
	.001	43	211.1	5	36	203.0	5	44	149.0	2
	.0001	130	347.6	6	127	349.7	7	152	249.5	2

To obtain numerical experience with ϵ-optimality the problem BMCF was investigated in Ruhe & Fruhwirth (1989). Doing this, at each iteration two minimum-cost flow problems must be solved. The relaxation code RELAXT of Bertsekas & Tseng (1988) was used as a subroutine.

In our computational tests we firstly fixed the cardinality of the solution set and report accuracy and cpu-time obtained from the application of ϵ-SANDWICH. This was done for fixed cardinalities s = 6,10, and 20 and is compared with the case that the whole set X_{eff} is calculated using the parametric method. The results of these tests are presented in Tables 5.7. and 5.8. Secondly, we fixed the accuracy to 0.1, 0.05, and 0.01 and report the cardinality of the calculated solution sets and the necessary cpu-times. Again, these results are compared with the exact case $\epsilon = 0$ in Tables 5.9. and 5.10.

To make the computational tests more attractive, we compared two versions of the ϵ-SANDWICH algorithm differing in the way to compute the two supporting lines at each iteration. We denote by EPS1 the ϵ-SANDWICH procedure with supporting lines whose gradients are defined by (33),(34). This is compared with a rule called angle trisection. The angle trisection rule also provides two gradients h^1 and h^2 which are obtained from trisecting the "outer" angle between the line segments of the lower approximating function l(t). The version of ϵ-SANDWICH using the angle trisection rule is called EPS2. Both versions are implemented in such a way that we always split the triangle causing the current ϵ-value. All tests performed include a comparison between EPS1 and EPS2.

For our numerical investigations we use five types of networks randomly generated by NETGEN (see Klingman, Napier & Stutz 1974). All the presented numerical results are average values of 20 instances generated for each type of network. We investigated networks of n = 600 vertices and m = 6000, 9000, and 12000 arcs and networks of n = 900 vertices with m = 9000 and 18000 arcs. The number of sources as well as the number of sinks was taken 1/3 of the number of vertices. The number of transshipment sources and transshipment sinks was taken one half of the number of sources and sinks, respectively. The arc capacities were chosen at random uniformly distributed in [500,5000]. The arc costs were chosen at random from the interval [1,1500].

All codes were written in Fortran. In Tables 5.8. and 5.10. we

Table 5.7. Numerical results for fixed cardinality: accuracy.

		s = 6		s = 10		s = 20		
n	m	EPS1	EPS2	EPS1	EPS2	EPS1	EPS2	Param.
600	6000	0.037	0.054	0.030	0.020	0.007	0.004	0
600	9000	0.057	0.087	0.041	0.030	0.009	0.006	0
600	12000	0.078	0.121	0.046	0.041	0.011	0.007	0
900	9000	0.037	0.054	0.031	0.020	0.007	0.004	0
900	18000	0.080	0.123	0.046	0.042	0.011	0.007	0

Table 5.8. Numerical results for fixed cardinality: cpu-seconds.

		s = 6		s = 10		s = 20		
n	m	EPS1	EPS2	EPS1	EPS2	EPS1	EPS2	Param.
600	6000	41	41	69	69	137	137	784
600	9000	56	57	94	93	187	187	1441
600	12000	74	75	124	124	245	249	2190
900	9000	75	77	126	127	254	256	1970
900	18000	139	142	236	235	464	470	5570

report average running times in cpu-seconds of a VAX 11/785 at the Graz University of Technology.

The computational results indicate that it is possible to handle the above stated difficulties of multi-criteria decision making in an appropriate way. By allowing a relative improvement of $\epsilon/(1+\epsilon)$ it is possible to construct solution sets of reasonable size already in the case of $\epsilon = 0.05$. Moreover, the measured cpu-time to calculate these sets is low when compared with the effort to calculate the complete set X_{eff}. Analogously, already for small sized solution sets of cardinality between 6 and 10 we obtain a "rather high" accuracy which should be sufficient in the case of applications. Concerning cpu-times we predict a further reduction when a strategy like the LTR-approach as suggested in Fruhwirth, Burkard & Rote (1989) is used.

Table 5.9. Numerical results for fixed accuracy: cardinality.

		$\epsilon = 0.1$		$\epsilon = 0.05$		$\epsilon = 0.01$		
n	m	EPS1	EPS2	EPS1	EPS2	EPS1	EPS2	Param.
600	6000	6	4	6	8	12	12	3547
600	9000	6	6	10	8	20	16	4559
600	12000	6	6	10	10	22	18	5309
900	9000	6	4	6	8	12	12	5876
900	18000	6	8	10	10	22	18	8821

Table 5.10. Numerical results for fixed accuracy: cpu-seconds.

		$\epsilon = 0.1$		$\epsilon = 0.05$		$\epsilon = 0.01$		
n	m	EPS1	EPS2	EPS1	EPS2	EPS1	EPS2	Param.
600	6000	41	27	41	55	82	83	784
600	9000	56	57	94	75	187	150	1441
600	12000	74	75	124	124	270	223	2198
900	9000	75	51	75	103	151	153	1970
900	18000	139	189	236	235	511	425	5570

From the point of view of real-world decision making the main advantage is that the decision maker can fix either the accuracy or the cardinality of the solution set in advance. From geometrical reasons we can argue that the calculated ϵ-optimal sets are well distributed in the objective space and may be considered representative of the complete set. Finally, we observe that there is no overall winner between the two versions EPS1 and EPS2.

5.8. An Application: Optimal Computer Realization of Linear Algorithms

Algorithms of linear structure are among those which are most frequently used in computer science. Numerical investigations concerning the implementation of linear algorithms indicate that the optimal implementation strategy highly depends on the performance of the computer being used. The difference in performance results from different values of parameters such as the size of the memory and the time necessary to execute an arithmetical or logical operation. In Groppen (1987), the problem of finding an optimal implementation variant for a given method is considered as a one-criterion problem combining two objectives:

(i) minimizing the time of computation subject to constraints on the size of the memory, and

(ii) minimizing the computer memory subject to constraints on the time of computation.

In Ruhe (1987), the two objectives are treated separately by means of a multicriteria approach. Due to the concrete circumstances, the decision maker can choose the suitable variant of implementation.

It is assumed that the consecutive states of a linear algorithm are described by a set of k-tuples. The different states may be represented by the set V of vertices of a directed graph $G = (V,A)$. An arc $(i,j) \in A$ indicates that there is a transition between the states i and j, i.e., the values characterizing the j-th state can be obtained from the values of the i-th state. The unique source 1 and sink n of G correspond to the final and terminal state of the algorithm. The linearity of the algorithm implies that G is acyclic. We have an one-to-one correspondence between the set of all paths in G connecting 1 an n and the set of all computer realizations of the algorithm under consideration. With each arc $(i,j) \in A$ a vector is associated describing the expenses required to calculate the information on state j from that of state i. For applying an optimal computer realization in respect to the time of computation and to the computer memory requirements, we use $t(i,j)$ respectively $m(i,j)$ to perform the transition between the states i and j. The set of all paths from 1 to n is abbreviated by **P**. The time $h(P)$ of a path $P \in \mathbf{P}$ is the sum of the times of the edges of P; the memory $m(P)$ is defined as the maximum of all memory requirements of the edges of P:

$$t(P) := \Sigma_{(i,j)\in P}\, t(i,j)$$
$$m(P) := \max \{m(i,j): (i,j) \in P\}.$$

Burkard, Krarup & Pruzan (1982) considered bicriteria (minisum and minimax) 0-1 programming problems. For a feasible set $F \subset \{0,1\}^n$ and nonnegative real functions c,d defined on F they investigated

(36) minisum: $\min \{c^T x: x \in F\}$

(37) minimax: $\min \{\max \{d_i x_i: 1 \le i \le n\}: x \in F\}$.

The bicriteria path problem with ojective functions (36) and (37) is called MSM.

Theorem 5.16. (Burkard, Krarup & Pruzan 1982)

(i) Each optimal solution of
CONVEX: $\min \{t \cdot c^T x + (1-t) \cdot \max_{1 \le i \le n} d_i \cdot x_i : x \in F\}$
with $0 < t < 1$ is an efficient solution of MSM, and

(ii) for any β not less than a nonnegative threshold value β^*, the union of the optimal solutions to
AUX: $\min \{t \cdot c^T x + (1-t) \cdot \max_{1 \le i \le n} d_i^{\beta} x_i : x \in F\}$
taken over all t, $0 < t < 1$, is identical to the set of all efficient solutions of MSM.

■

For the solution of MSM, we can use an algorithm easier than the approach motivated by the above theorem. Let $f(P) = (t(P), m(P))$ be the vector of the two objective functions. The original problem

$$(38) \quad \min{}^* \{f(P) : P \in \mathbf{P}\}$$

defined on $G = (V,A)$ is decomposed into a sequence of subproblems on $G^q = (V,A^q)$; $q = 1,\ldots,h$ with

$$A^q := A_1 + A_2 + \ldots + A_q \; ; \; q = 1,\ldots,h,$$
$$A_q := \{(i,j) \in A: m(i,j) = \alpha_q\} \text{ and } \alpha_1 < \alpha_2 < \ldots < \alpha_h .$$

Let $\mathbf{P}^q$ be the set of all paths from 1 to n in the subgraph G^q. Then we have to solve consecutively path problems of the form:

$$ASP^q: \min \{t(P): P \in \mathbf{P}^q\}.$$

Assume P_{eff} be the set of efficient solutions of (38).

Theorem 5.17.
Let P^q and P^{q-1} be optimal solutions of ASP^q respectively ASP^{q-1}. Then the following is valid for $q = 2,\ldots,h$:

(i) $t(P^q) \le t(P^{q-1})$

(ii) P^q is an optimal solution also for ASP^{q-1} iff $P^q \cap A_q = \emptyset$

(iii) $t(P^q) < t(P^{q-1})$ implies $P^q \in P_{eff}$.

Proof:
$A^{q-1} \subset A^q$ implies that $t(P^{q-1}) \ge t(P^q)$. If P^q is an optimal solution of ASP^{q-1}, then $P^q \in \mathbf{P}^{q-1}$. Thus it holds $P^q \cap A^q = \emptyset$. Conversely, from $P^q \cap A^q = \emptyset$ and (i) it follows that P^q is optimal for ASP^{q-1}.

To show the efficiency of P^q consider a path $P \in \mathbf{P}$ and assume that $m(P) < m(P^q)$. Consequently, $P \in \mathbf{P}^r$ for an index r such that $r < q$. From (i) it follows $t(P) \geq t(P^q)$. However, the equality is imposible because of $t(P^q) > t(P^{q-1})$. In a similar way we prove that the existence of a path P' with $t(P') < t(P^q)$ implies that $m(P') > m(P^q)$. In fact, our assumption implies that $P' \in \mathbf{P}^r$ with $r > q$. Consequently, $m(P') > m(P^q)$ since otherwise there would be a contradiction to the optimality of P^q with respect to ASP^q.

■

From the above theorem we conclude that for $t(P^r) = t(P^{\lfloor r/2 \rfloor})$ the subproblem defined by the sets $A_{\lfloor r/2+1 \rfloor},\ldots,A_r$ can be neglected from the further considerations. Applying this test repeatedly, we define a procedure called CONQUER. Without loss of generality we assume that h $= 2^k$ for an integer k. CONQUER uses DIVIDE and SHORTPATH as a subroutine. For SHORTPATH, one of the efficient procedures existing for this class of problems (for a survey it is referred to Gallo & Pallottino 1989) can be taken.

Although only shortest path problems are considered here, the results and methods are valid for other network optimization problems, too. The common feature of these models is that the feasible solutions of the problem are described by certain subgraphs with special properties. For instance, this may be paths connecting two vertices, spanning trees or matchings. In all these cases we can use the same idea to find efficient solutions via a polynomial algorithm. From the viewpoint of computational complexity, this result is of interest because it is known that bicriteria problems with two additive functions are NP-complete in the case of shortest paths (Garey & Johnson 1979) and of minimum spanning trees (Aggarwal, Aneja & Nair 1982).

procedure CONQUER
begin
  $P_{eff} := \emptyset$
  $A_0 := \emptyset$
  DIVIDE(A_p, A_0)
  for j := 1 **to** p-1 **do**
    if $t(P^{p-j+1}) < t(P^{p-j})$ **then** $P_{eff} := P_{eff} + \{P^{p-j+1}\}$
  if $P^1 \neq \emptyset$ **then** $P_{eff} := P_{eff} + \{P^1\}$
end

```
procedure DIVIDE(A_r,A_s)
begin
C Assume r < s
  t:= r + ⌊(s-r)/2⌋
  P^r := SHORTPATH(A_r)
  if ⌊(s-r)/2⌋ > 1 and t(P^s) < t(P^t) then DIVIDE(A_r,A_t)
  if ⌊(s-r)/2⌋ > 1 then DIVIDE(A_t,A_s)
end
```

§6 PARAMETRIC FLOWS

6.1. Motivation and Fundamental Results

In parametric programming the behaviour of functions, algorithms or solutions in dependence of modifications in the problem data are investigated. The importance of this kind of analysis is primarily due to occurrences of long sequences of problem instances, where each instance differs from the others by small modifications of the problem data. Gusfield (1980) gives several reasons for such long sequences:

1. Inexact data or model: When the problem data are not known exactly, and optimization models rarely capture all the problem constraints and relationships, then it is useful to solve the optimization model with different choices of data, or different choices of the model.

2. Changing data or model: Even for exact model and data, the problem may be dynamic and changing. Then a sequence of successive instances of the problem must be solved, where each instance differs from the others by the modification of some parts of the problem data.

3. Conflicting objectives: As described in Chapter 5, in most applications there is more than only one objective function. Then in case of conflicting objectives there may be different solutions which are "good" with respect to the different objectives. Parametric programming reflects the sensitivity of the solutions related to different weightings of the objective functions.

4. Heuristics for NP-hard problems: Many NP-hard optimization problems are solved heuristically. One approach is to tackle the problem by embedding fast algorithms for related problems in the inner loop of a larger program and calling the fast algorithms repeatedly with successively modified problem data. Interactive optimization, Branch & Bound and Lagrangean relaxation are the most important representatives of this class of solution methods.

Before the design of special algorithms for parametric flow problems we give some general approaches for the problem

PP $\quad F(t) = \min \{c^T x + t \cdot d^T x : x \in \mathcal{X},\ t \in [0,T]\}$

or its dual program having parametric right-hand sides. From parametric linear programming (compare Nozicka et al. 1974) the following result is well known:

Theorem 6.1.
F(t) is a piecewise linear and concave function. For two breakpoints at $t_k < t_{k+1}$ and an optimal solution x^* for t^*: $t_k < t^* < t_{k+1}$, x^* is optimal over the entire closed interval $[t_k, t_{k+1}]$.
■

Eisner & Severanne (1976) developed a general method for finding all the breakpoints of F(t). The basic operation is to calculate $F(t^*)$ for a fixed parameter t^*. With the two solutions x_0 and x_T the unique value t1 is determined where $c^Tx_0 + t1 \cdot d^Tx_0 = c^Tx_T + t1 \cdot d^Tx_T$. Subsequently, F(t1) is determined. If $F(t1) < c^Tx_0 + t1 \cdot d^Tx_0$ then there will be two recursive calls of the procedure for the intervals [0,t1] and [t1,T]. Doing in this way, the procedure iteratively partitions the given interval into subintervals as long as two considered breakpoints are not proven to be adjacent. For any piecewise linear function F(t) with k breakpoints the above described method needs 2k - 1 evaluations of F(t).

The optimal objective value function can be computed in a vertical fashion or "piece-by-piece". The idea is to assume that the optimality function $x^*(t)$ is known for $t \in [0,t_k]$ and $t_k < T$. Then we have to answer the two related questions:

(i) Which flow change dx ensures the optimality of $x^*(t_k + \epsilon) := x^*(t_k) + \epsilon \cdot dx$ for sufficiently small $\epsilon > 0$?

(ii) Which is the maximum length dT such that $x^*(t_k) + dt \cdot dx$ is optimal in $[t_k, t_k + dT]$?

This approach was applied for the parametric maximum flow problem in Ruhe (1985b) and is described in Section 6.3. This includes the results of numerical investigations of Klinz (1989) comparing the vertical algorithm with two other methods for solving the parametric maximum flow problem.

Hamacher & Foulds (1989) investigated an approximation approach for calculating $x^*(t)$ for all t in [0,T]. The idea is to determine in each iteration an improvement dx(t) defined on the whole parameter interval. dx(t) is a continuous and piecewise linear function. According to its geometrical interpretation, this is called a horizontal approach. The horizontal approach is applied to maximum flows in generalized networks in Section 6.4. Vertical and horizontal approach are illustrated in Figure 6.1.

In general, dual methods seem to be suited for reoptimization in the case that primal feasibility is destroyed and dual feasibility is maintained. The parametric dual simplex method is investigated in Chapter 6.5. to handle parametric changes in the minimum-cost flow problem.

The approximation approach of Burkard, Hamacher & Rote (1987) described in Section 5.5. can be applied to approximate the optimal value function F(t). The main advantage of this method is to have an estimate of the error between the actually calculated and the optimal solution.

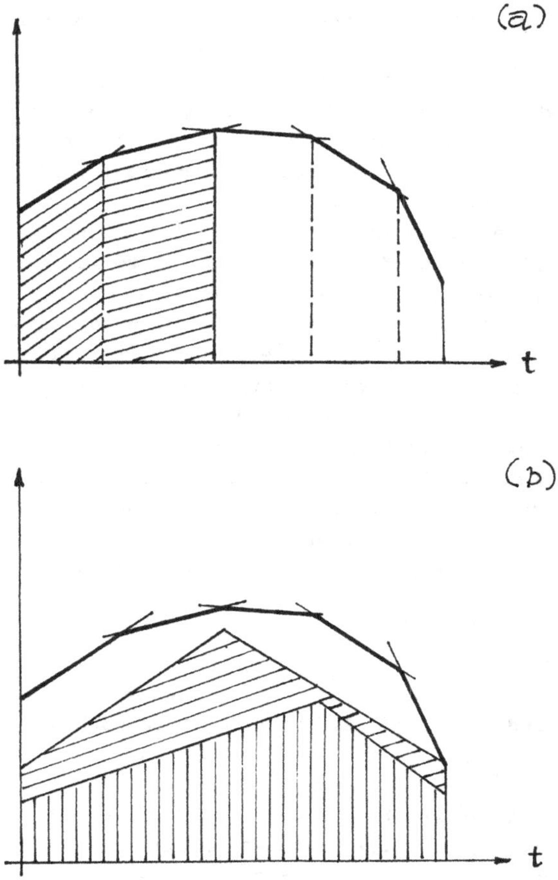

Figure 6.1. (a) The vertical approach.
(b) The horizontal approach.

6.2. The Number of Breakpoints for Parametric Flow Problems

A *breakpoint* is a place where the slope of the optimal value function of a parametric optimization problem is changing. One approach to express complexity of a parametric problem is to count the number of breakpoints. This provides a bound both on the amount of computation and memory which is necessary in the worst case. Murty (1980) has shown that in the general case of parametric linear programming, the number of breakpoints in the optimal value function exponentially depends on the number of variables. Charstensen (1983) proved similar results for parametric integer programming and parametric minimum-cost flows with the parameter in the cost function. In the case of the parametric shortest (directed) path problem an example with $n^{\log n} - 1$ breakpoints was given by Charstensen (1984). Gusfield and Irving (1989) enhanced the classical Gale-Shapley stable marriage approach and considered parametric stable marriages. They derived a bound of n(n-1)/2 breakpoints for the corresponding optimal value function.

For the parametric maximum flow problem with linear capacity functions Cap(i,j;t) := cap(i,j) + t·pcap(i,j) for all arcs (i,j) of A and a parameter t varying in the interval [0,T] Gallo, Grigoriadis & Tarjan (1989) considered a special case in which the capacities of the arcs leaving the source are nondecreasing and the arcs entering the sink are nonincreasing functions of the parameter t, while the capacities of all other arcs are constant. They proved that the number of breakpoints is not greater than n in this case. Martel (1987) compared phase and non-phase maximum flow algorithms and used a special parametric network shown in Figure 6.2. The only condition is that for the capacities of the corresponding arcs

$$0 < cap(n-1,n-2) < cap(n-2,n-3) < \ldots < cap(3,2)$$

must be satisfied.

Lemma 6.1. (Martel 1987)
The parametric maximum flow problem of Figure 6.2. with parameter interval [0,T] and $T \geq cap(3,2)$ has n - 3 breakpoints.

Proof:
The minimum cuts are given by
{1},{1,n-1},{1,n-1,n-2},....,{1,n-1,n-2,...,2}. This results in breakpoints at t = cap(n-1,n-2),cap(n-2,n-3),....,cap(3,2), respectively. ■

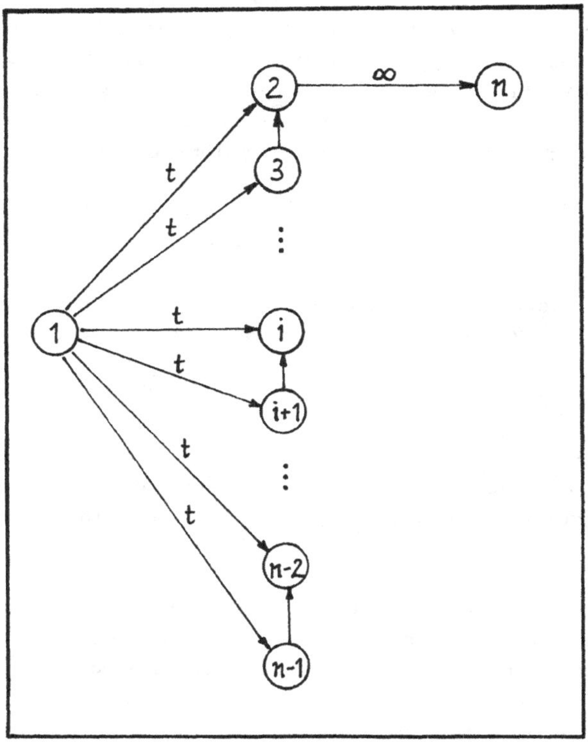

Figure 6.2. The network of Martel (1987).

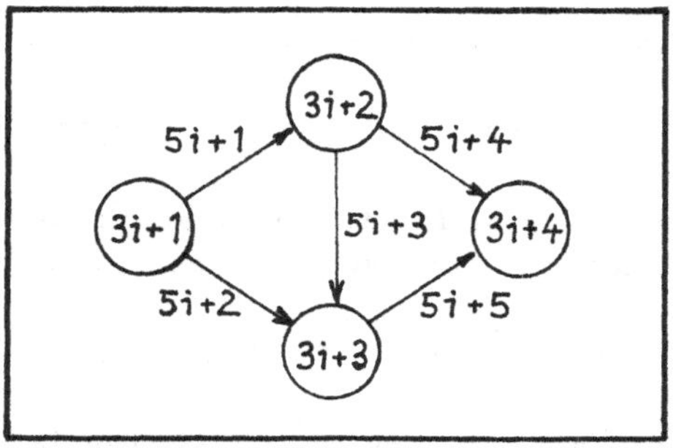

Figure 6.3. Subgraphs G(i); i = 0,...,k for the network $N_{4/3n}(k)$.

The question to determine general upper bounds for parametric maximum flows in pure networks is an open problem. Klinz (1989) defined a class of networks $N_{4/3n}(k)$ having $n = 4 + 3k$ vertices and $m = 5(k + 1)$ arcs for all $k \geq 0$, $k \in \mathbf{Z}$. $N_{4/3n}(k)$ is composed of $k + 1$ subgraphs of the form shown in Figure 6.3. Using arc numbers i as introduced in the figure, the capacity functions Cap(i;t) are defined as

$$(1)\quad Cap(i;t) := \begin{cases} 10 + 20t & \text{for } i \equiv 1 \bmod 5 \\ 10(7i^2 + 6i + 1) + (w - 35i - 20)t & \text{for } i \equiv 2 \bmod 5 \\ 20 & \text{for } i \equiv 3 \bmod 5 \\ 10(6i + 3) + 5t & \text{for } i \equiv 4 \bmod 5 \\ 10(7i^2 + 2i) + (w - 35i - 20)t & \text{for } i \equiv 0 \bmod 5 \end{cases}$$

where w is a sufficiently large number.

Lemma 6.2. (Klinz 1989)
The parametric flow problem defined by the subgraphs of Figure 6.3. and capacity function (1) has $(4n - 7)/3$ breakpoints in the parameter interval $[0,4k + 4]$.

■

We investigate the complexity of the parametric maximum flow problem in generalized networks and present a pathological instance of the problem causing an exponential number of breakpoints.

The dependence of t for the flow on the arc $(i,j) \in A$ is denoted by $x(i,j;t)$. The polyhedron $\mathbf{X}_g$ in the formulation of MGF is replaced by a sequence of polyhedrons $\mathbf{X}_g(t)$ defined by

$$\Sigma_{(j,k)\in A}\, g(j,k)\cdot x(j,k;t) - \Sigma_{(i,j)\in A}\, x(i,j;t) = 0 \quad \text{for all } j \in V-\{1,n\}$$

$$0 \leq x(i,j;t) \leq cap(i,j) + t\cdot pcap(i,j) \qquad \text{for all } (i,j) \in A$$

where pcap: $A \longmapsto \mathbf{R}$ is used for the parametric part of the upper capacity constraints. Now the problem may be stated as:

PMGF opt $\{ x_n(t) : x(t) \in \mathbf{X}_g(t)$ for all $t \in [0,T]\}$.

Our bad generalized network proving the exponential growth of the number of breakpoints in dependence of #(V) has two ingredients: the topological structure of the graph G_n shown in Figure 5.1., and special assignments for the capacities and the multipliers. For each arc $(i,j) \in A_n$ we define

(2) $g(i,j) := 2^{-\Gamma(i,j)}$ with

$$\Gamma(i,j) := \begin{cases} \infty & \text{for } (i,j) = (0,\underline{0}) \\ 0 & \text{for } (i,j) = (0,k),(\underline{k},\underline{0});\ k = 1,\dots,n \\ 2^{\max\{i,j\}-1} - 1 & \text{otherwise.} \end{cases}$$

For brevity, we use $\sigma(k) := \lceil \Gamma(P_k)/2 \rceil$ and $\mu(k) := \lfloor \Gamma(P_k)/2 \rfloor$.

The capacities of our pathological example are:

$$\text{(3)}\quad cap(i,j) := \begin{cases} 1 & \text{for } (i,j) = (0,1),(\underline{2},\underline{0}) \\ 2 & \text{for } (i,j) = (\underline{1},\underline{0}) \\ 8 & \text{for } (i,j) = (0,2) \\ \Sigma_{h \in I(k)}\ 2^{\sigma(h)} & \text{for } (i,j) = (0,k);\ k = 3,\dots,n \\ \Sigma_{h \in I(k)}\ 2^{-\mu(h)} & \text{for } (i,j) = (\underline{k},\underline{0});\ k = 3,\dots,n \\ \infty & \text{otherwise} \end{cases}$$

where $I(k) := \{r_{k-1} + 2,\dots,r_k + 1\}$ and $\sigma(r_n + 1) = \mu(r_n + 1) := 2^{n-2}$.

We restrict ourselves to the case that a parametric amount of flow is leaving the source. By introducing an additional arc $(0^*,0)$ and using $cap(0^*,0) = 0$ we define

$$pcap(i,j) := \begin{cases} 1 & \text{for } (i,j) = (0^*,0) \\ 0 & \text{otherwise} \end{cases} .$$

This is a special case of PMGF. Based on the paths $P_1,\dots,P_{r(n)}$, generalized flows $x^1,\dots,x^{r(n)}$ are defined by

$$\text{(4)}\quad x^k(i,j) := \begin{cases} (-)2^{\sigma(k)} \cdot g(P_k[1,i]) & \text{for } (i,j) \in P_{k+}\ (\in P_{k-}) \\ 0 & \text{for } (i,j) \notin P_k \end{cases}$$

for $k = 1,\dots,r(n)$. In Table 6.1. the flow values $x^k(i,j)$; $(i,j) \in A_3$; $k = 1,\dots,8$ related to the paths P_k of Figure 5.2. are presented.

The following result is a generalization of Lemma 5.1.

Lemma 6.3. (Ruhe 1988b)
Let P_k, P_l be a pair of augmenting paths in G_n, such that $P_l = [0,n,\overleftarrow{P_k},\underline{n},\underline{0}]$. Then $x^k(i,j) + x^l(i,j) = 0$ for all arcs (i,j) contained both in P_k and P_l.

■

Table 6.1. Generalized flows x^k; k=1,...,8 related to the paths of Figure 5.2.

k	(0,1)	(0,2)	(0,3)	$(1,\underline{1})$	$(1,\underline{2})$	$(1,\underline{3})$	$(2,\underline{1})$	$(2,\underline{3})$	$(3,\underline{1})$	$(3,\underline{2})$	$(\underline{1},\underline{0})$	$(\underline{2},\underline{0})$	$(\underline{3},\underline{0})$
1	1			1							1		
2		2					2				1		
3		2		-1	1		2					1/2	
4		4						4					1/2
5			4							4		1/2	
6			4	1	-1		-2	2		4			1/4
7			8				-2	2	8				1/4
8			8	-1		1			8				1/8

In the special case of acyclic graphs, the solution of PMGF for a fixed parameter t can be obtained from the consecutive search for flow augmenting paths of maximum multiplicative weight. Before performing this approach with respect to the sequence $P_1,\ldots,P_{r(n)}$, the flows y^0 and

$$y^k := \sum_{j=1}^{k} x^j \quad \text{for } k = 1,\ldots,r(n)$$

are defined. We assume that $X_g(t) \neq \emptyset$ for the whole parameter interval [0,T].

Lemma 6.4. (Ruhe 1988b)
For each of the flows x^k; k=1,...,r(n) as defined in (4), the following propositions are valid:

(i) x^k is a feasible and maximal flow in $G(y^{k-1})$, and

(ii) the augmentation is along a multiplicative longest path among all paths between 1 and n in $G(y^{k-1})$.

■

Without loss of generality we assume that the index set described by Lemma 5.3. is given by $I_n = \{1,\ldots,s(n)\}$.

Theorem 6.2.

The instance of PMGF defined by G_n, the multipliers (2), and the capacities given by (3) cause an exponential number $q = 2^n - 2$ of breakpoints having coordinates

$$(t, \#(x_n(t))) = \left(\sum_{i=1}^{k} 2^{\sigma(i)} , \sum_{i=1}^{k} 2^{-\mu(i)} \right); \; k = 1,\ldots,s(n)$$

in the optimal value function.

Proof:

From Lemma 5.2. the existence of paths P_k, $k \in I_n$ is derived such that $g(P_i) > g(P_j)$ for all $i,j \in I_n$ with $i < j$ because of (2). For

$t = \sum_{i=1}^{k} 2^{\sigma(i)}$, $k = 1,\ldots r(n)$, the optimal flow is given by

sending $\#(x_1(t)) = t$ units of flow along the paths $P_1,\ldots P_k$. This results in $\#(x_n(t))$ units of flow reaching the sink with

$$\#(x_n(t)) = \sum_{i=1}^{k} 2^{\mu(i)} \quad \text{and} \quad t = \sum_{i=1}^{k} 2^{\sigma(i)} , \; k = 1,\ldots,r(n).$$

That means, we have obtained a piecewise linear function defined in the intervals

$$[0,2^{\sigma(1)}], \; [2^{\sigma(1)},2^{\sigma(1)} + 2^{\sigma(2)}],\ldots, \; [\sum_{i=1}^{r(n)-1} 2^{\sigma(i)}, \; \sum_{i=1}^{r(n)} 2^{\sigma(i)}].$$

For the k-th interval there is a slope of

$$d(k) := 2^{-\mu(k)} / 2^{\sigma(k)} = 2^{-\Gamma(Pk)} = g(P_k); \; k = 1,\ldots,r(n).$$

Using Lemma 5.3., we get $2^n - 1$ different slopes and $q = 2^n - 2$ breakpoints as given above.

∎

6.3. Vertical Algorithm for the Parametric Maximum Flow Problem

We consider the parametric maximum flow problem with linear capacity functions $Cap(i,j;t) := cap(i,j) + t \cdot pcap(i,j)$ for all arcs (i,j) of A and a parameter t varying in the interval $[0,T]$. The polyhedron $\mathbf{X}$ in the formulation of MF is replaced by a sequence of polyhedrons $\mathbf{X}(t)$ for all $t \in [0,T]$. Each $\mathbf{X}(t)$ is described by

$$\sum_{(j,k)\in A} x(j,k;t) - \sum_{(i,j)\in A} x(i,j;t) = 0 \quad \text{for all } j \in V - \{1,n\}$$
$$0 \le x(i,j;t) \le Cap(i,j;t) \quad \text{for all } (i,j) \in A.$$

We assume that $\mathbf{X}(t) \neq \emptyset$ for the whole parameter interval.

PMF $\max \{\#(x(t)): x(t) \in \mathbf{X}(t) \text{ for all } t \in [0,T]\}$.

Gallo, Grigoriadis & Tarjan (1989) considered a special case of PMF in which the capacities of the arcs leaving the source are nondecreasing and the arcs entering the sink are nonincreasing functions of the parameter t, while the capacities of all other arcs are constant. They extend the preflow algorithm of Goldberg & Tarjan (1986) to this more general class of problems. Their parametric preflow algorithm is only a constant factor greater than the time bound to solve a non-parametric problem of the same size and runs in $O(n \cdot m \cdot \log(n^2/m))$ time. The minimum cuts produced by the algorithm have a nesting property, i.e., for a given on-line sequence of parameter values $t_1 < t_2 < \ldots < t_q$ the parametric preflow algorithm computes cuts (X_1,X_1^*), $(X_2,X_2^*),\ldots,(X_q,X_q^*)$ such that $X_1 \subset X_2 \subset \ldots \subset X_q$. The main idea in their algorithm is to modify a maximum flow x computed for a fixed parameter $t_k \in [0,T]$ with the preflow algorithm to the preflow x^* with

$$x^*(i,j) := \begin{cases} \max \{Cap(i,j;t_{k+1}), x(i,j;t_k)\} & \text{for } (i,j)=(1,h) \text{ and } d(h)< n \\ \min \{Cap(i,j;t_{k+1}), x(i,j;t_k)\} & \text{for } (i,j)=(h,n) \\ x(i,j;t_k) & \text{otherwise.} \end{cases}$$

A maximum flow and a minimum cut for the next parameter t_{k+1} can be computed by applying the preflow algorithm with the modified x^* and the current labeling d. Among the applications the authors investigate different formulations of flow sharing and zero-one fractional programming problems.

We use the abbreviations $X_{max}(t)$ and $C_{min}(t)$ for the set of maximum flows respectively minimum cuts for PMF with fixed parameter t.

The idea of the vertical implementation is to assume that the optimality function $x^*(t)$ is known for $t \in [0,t_k]$, $t_k < T$. Then we determine a flow change dx ensuring the optimality of $x^*(t_k+\epsilon) := x^*(t_k) + \epsilon \cdot dx$ for $t > t_k$. Therefore, the following subproblem is investigated:

MF(x): $\max \{\#(dx) : \#(dx) = \Sigma_{(1,j)\in A} dx(1,j)$

$$\Sigma_{(j,k)\in A} dx(j,k) - \Sigma_{(i,j)\in A} dx(i,j) = 0 \quad \text{for all } j \in V - \{1,n\}$$
$$l(i,j) \le dx(i,j) \le u(i,j) \quad \text{for all } (i,j) \in A \quad \}$$

with capacity bounds defined as:

$$l(i,j) := \begin{cases} 0 & \text{for } x(i,j) = 0 \\ -\infty & \text{for } x(i,j) > 0 \end{cases}$$

and

$$u(i,j) := \begin{cases} pcap(i,j) & \text{for } x(i,j) = Cap(i,j;t_k) \\ \infty & \text{for } x(i,j) < Cap(i,j;t_k). \end{cases}$$

Each subproblem is a max-flow problem again and can be solved via a strongly polynomial algorithm.

Let dx be a solution of MF(x). To determine the stability interval of the so determined solution we need a partition of the set of arcs A into disjoint subsets A(k); k = 1,...,4 with

$$A(1) := \{(i,j) \in A: dx(i,j) \ge 0 \ \& \ dx(i,j) > pcap(i,j)\}$$
$$A(2) := \{(i,j) \in A: dx(i,j) < 0 \ \& \ dx(i,j) \le pcap(i,j)\}$$
$$A(3) := \{(i,j) \in A: dx(i,j) < 0 \ \& \ dx(i,j) > pcap(i,j)\}$$
$$A(4) := \{(i,j) \in A: dx(i,j) \ge 0 \ \& \ dx(i,j) \le pcap(i,j)\}.$$

From the definition of the subsets it follows that each arc (i,j) of A is contained in exactly one subset. The sets A(1), A(2), and A(3) may result in restrictions for the length dT of the interval:

$$Q_1(i,j) = (Cap(i,j;t_k)-x(i,j;t_k))/(dx(i,j)-pcap(i,j)) \quad \text{for } (i,j) \in A(1)$$
$$Q_2(i,j) = -x(i,j;t_k)/dx(i,j) \quad \text{for } (i,j) \in A(2)$$
$$Q_3(i,j) = \min \{Q_1(i,j), Q_2(i,j)\} \quad \text{for } (i,j) \in A(3)$$

Let be $dT_k := \min \{Q_k(i,j): (i,j) \in A(k)\}$ for k = 1,2,3.

Theorem 6.3.
Let be $x(t_k) \in X_{max}(t_k)$ for $t_k < T$ and dx a solution of $MF(x(t_k))$.
Then (i) $\#(dx) = \min \{\Sigma_{(i,j)\in\delta+(X)} pcap(i,j): (X,X^*) \in C_{min}(t_k)\}$
(ii) $x(t_k) + dt\cdot dx \in X_{max}(t_k + dt)$ for $0 \le dt \le dT$ with $dT := \min \{T - t_k, dT_1, dT_2, dT_3\}$.

$dT := \min \{T - t_k, dT_1, dT_2, dT_3\}$.

Proof:

The capacity $Cap(X,X^*)$ of a cut (X,X^*) with non-zero lower bounds $l \neq 0$ and upper capacity bounds $u \geq l$ is given by

$$Cap(X,X^*) = \Sigma_{(i,j)\in\delta+(X)} u(i,j) - \Sigma_{(i,j)\in\delta-(X)} l(i,j).$$

In the case of $MF(x(t_k))$ this results in

$$\begin{aligned} Cap(X,X^*) = &\Sigma_{(i,j)\in\delta+(X):x(i,j;tk)=Cap(i,j;tk)} pcap(i,j) \\ &+ \Sigma_{(i,j)\in\delta+(X):x(i,j;tk) < Cap(i,j;tk)} \infty \\ &- \Sigma_{(i,j)\in\delta-(X):x(i,j;tk) > 0} (-\infty). \end{aligned}$$

As a consequence of the max-flow min-cut theorem it holds

$(i,j) \in \delta^+(X)$ implies $x(i,j;t_k) = Cap(i,j;t_k)$

$(i,j) \in \delta^-(X)$ implies $x(i,j;t_k) = 0$

for all cuts $(X,X^*) \in C_{min}(t_k)$ and $x(t_k) \in X_{max}(t_k)$. This implies our first proposition:

$$\begin{aligned} \#(dx) &= \min \{Cap(X,X^*): (X,X^*) \in C(t_k)\} \\ &= \min \{\Sigma_{(i,j)\in\delta+(X)} pcap(i,j) : (X,X^*) \in C_{min}(t_k)\}. \end{aligned}$$

For the second part we firstly show the feasibility in each of the four cases $(i,j) \in A(k)$; $k = 1,\ldots,4$:

a) $(i,j) \in A(1)$

The validity of $0 \leq dt \leq dT \leq dT_1 \leq Q_1(i,j)$ results in

$$\begin{aligned} &x(i,j;t_k) + dt\cdot dx(i,j) \leq x(i,j;t_k) + Q_1(i,j)\cdot dx(i,j) \\ &= x(i,j;t_k) + [(Cap(i,j;t_k) - x(i,j;t_k))/(dx(i,j)-pcap(i,j)]\cdot dx(i,j). \end{aligned}$$

For $(i,j) \in A(1)$ it holds $dx(i,j) \geq 0$ and $dx(i,j) > pcap(i,j)$. In the case that $dx(i,j) > 0$ this results in the fulfillment of the upper bounds at $t = t_k + dt$.

For $dx(i,j) = 0$ the feasibility follows from

$$dt \leq (Cap(i,j;t_k) - x(i,j;t_k))/(dx(i,j)-pcap(i,j)).$$

Concerning the lower bounds the non-negativity of $dx(i,j)$, dt, and $x(i,j;t_k)$ ensures that $x(i,j;t_k) + dt\cdot dx(i,j) \geq 0$.

b) $(i,j) \in A(2)$

$Q_2(i,j) = -x(i,j;t_k)/dx(i,j) \geq dt \geq 0$ and $dx(i,j) < 0$ leads to $x(i,j;t_k) + dt\cdot dx(i,j) \geq 0$. The fulfillment of the upper bounds is a consequence of $x(i,j;t_k) \leq Cap(i,j;t_k)$ and $dx(i,j) \leq pcap(i,j)$.

c) $(i,j) \in A(3)$

Because of $Q_3(i,j) = \min \{Q_1(i,j), Q_2(i,j)\}$, the inequalities of the cases a) and b) are valid.

d) $(i,j) \in A(4)$

In this case, the feasibility of $x(t_k) + dt\cdot dx$ is a consequence of $0 \leq dx(i,j) \leq pcap(i,j)$, $dt \geq 0$, and the feasibility of $x(i,j;t_k)$.

To verify optimality we show that $x + dt\cdot dx$ satisfies the max-flow

$$\begin{aligned}
&\max \{\#(x) : x \in \mathbf{X}(t_k+dt)\} \\
&= \#(x(t_k) + dt \cdot dx) \\
&= \#(x(t_k)) + dt \cdot \#(dx) \\
&= \min \{Cap(X,X^*) : (X,X^*) \in C(t_k)\} + dt \cdot \min \{\Sigma_{(i,j)\in\delta+(X)} pcap(i,j) : (X,X^*) \in C_{min}(t_k)\} \\
&= \min \{Cap(X,X^*) : (X,X^*) \in C(t_k+dt)\}
\end{aligned}$$

■

From the iterative application of the above theorem an algorithm for PMF follows. We give a high-level description using the solution of the max-flow problem MF(x) as a subroutine.

```
procedure VERT-PMF
begin
  t := 0
  x(0) := 0
  while t < T do
  begin
    dx := MF(x)
    for k := 1 to 3 do dT_k := min {Q_k(i,j): (i,j) ∈ A(k)}
    dT := min {T-t,dT_1,dT_2,dT_3}
    t := t + dT
  end
return
end
```

We illustrate the algorithm by a numerical example and consider the graph of Figure 6.4. The parameter interval for t is [0,1]. For t = 0 the flow x^0 shown in Figure 6.5.(a) is maximum. The min-cut is described by the set X_0 with $X_0 = \{1,2,3,5\}$ and $Cap(X_0,X_0^*) = 8 = \#(x^0)$. From there we obtain a subproblem $MF(x^0)$. The corresponding maximal flow is given in Figure 6.5.(b).

Because of $C_{min}(0) = \{X_0\}$, for the optimal solution dx^0 of $MF(x^0)$ it holds $\#(dx^0) = 5$. To determine the parameter interval where x^0 is optimal, we have to consider the sets A(k); k = 1,...,4:

$$\begin{aligned}
A(1) &= \{(1,3),(4,6),(4,7),(6,7)\}; \\
A(2) &= \{(5,7)\}; \\
A(3) &= \{(2,5)\}; \\
A(4) &= \{(1,2),(1,4),(2,3),(3,6)\}.
\end{aligned}$$

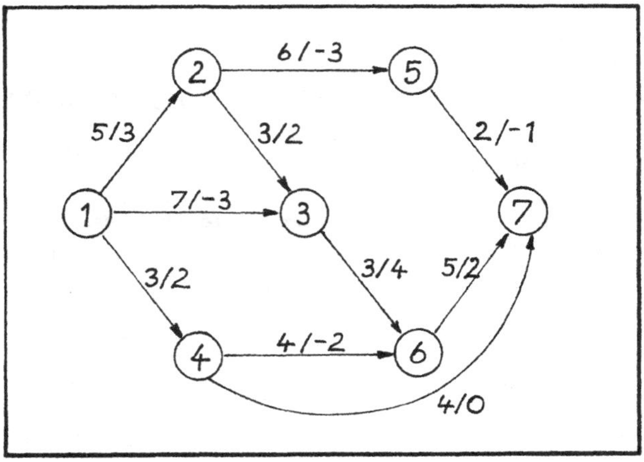

Figure 6.4. Graph G = (V,A) with cap(i,j), pcap(i,j) for (i,j) $\in$ A.

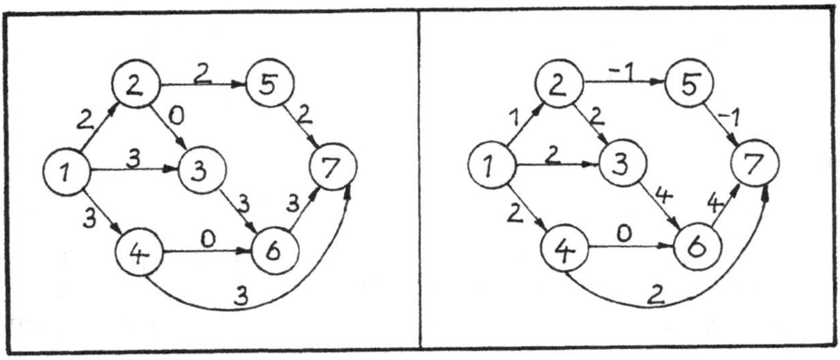

(a) (b)

Figure 6.5. (a) Maximum flow x^0 for t = 0.
(b) Maximum flow of MF(x^0).

This implies $T_1 = \min \{4/5,\ 4/2,\ 1/2,\ 2/2\} = 1/2$,

$T_2 = 2$, $T_3 = \min \{4/2,\ 2\} = 2$,

$T = \min \{1, 1/2, 2, 2\} = 1/2$.

That means $x^1(t) := x^0 + t \cdot dx^0$ is optimal for $t \in [0,1/2]$. The steps described above are repeated now on the foundation of x^1. We have summarized the corresponding numerical results in Table 6.2.

Table 6.2. Solution of the parametric maximum flow problem given in Figure 6.4.

(i,j)	(1,2)	(1,3)	(1,4)	(2,3)	(2,5)	(3,6)	(4,6)	(4,7)	(5,7)	(6,7)
$x^1(i,j)$	5/2	4	4	1	3/2	5	0	4	3/2	5
$l^1(i,j)$	-M	-M	-M	-M	-M	-M	0	-M	-M	-M
$u^1(i,j)$	M	M	2	M	M	4	M	0	2	M
$dx^1(i,j)$	3	0	2	4	-1	4	2	0	-1	6
$x^2(i,j)$	13/4	4	9/2	2	5/4	6	1/2	4	5/4	13/2
$l^2(i,j)$	-M	-M	-M	-M	-M	-M	-M	-M	-M	-M
$u^2(i,j)$	M	M	2	M	M	4	M	0	2	2
$dx^2(i,j)$	1	0	0	2	-1	2	0	0	-1	2
$x^3(i,j)$	7/2	4	9/2	5/2	1	13/2	1/2	4	1	7

On the foundation of dx^1 we obtain the sets of arcs:

$A(1) = \{(1,3),(2,3),(4,6),(6,7)\}$,

$A(2) = \{(5,7)\}$, $A(3) = \{(2,5)\}$,

$A(4) = \{(1,2),(1,4),(3,6),(4,7)\}$.

Consequently, $dT = dT_1 = Q_1(6,7) = 1/4$. In the same way, on the foundation of dx^2, $dT = dT_1 = Q_1(1,3) = 1/4$ is determined. Now the optimality function x(t) is calculated as

$$x(t) = \begin{cases} x^0 + t\cdot dx^0 & \text{for } t \in [0,1/2] \\ x^1 + (t-1/2)\cdot dx^1 & \text{for } t \in [1/2,3/4] \\ x^2 + (t-3/4)\cdot dx^2 & \text{for } t \in [3/4,1] \end{cases} .$$

Klinz (1989) made numerical investigations to compare VERT-PMF with the parametric dual simplex method and the sandwich algorithm. For that reason, the networks $N_{4/3n}(k)$ based on Figure 6.3. were used. The characteristics of the test examples are given in Table 6.3.

In Table 6.4., some of the numerical results comparing VERT-PMF and the parametric dual network simplex method are summarized. The main reason for the superiority of the dual method is its ability to solve the consecutive problems more quickly using reoptimization. Another important feature of VERT-PMF is the large number of degenerate iterations,i.e., flow changes which do not lead to a new breakpoint. Klinz (1989) presented a pathological example where VERT-PMF needs an infinite number of iterations if it is used without modifications.

Table 6.3. Characteristics of the test examples of Klinz (1989).

#	k	n	m	θ2(PMF)
1	25	79	130	103
2	50	154	255	203
3	75	229	380	303
4	100	304	505	403
5	250	754	1255	1003
6	500	1504	2505	2003
7	750	2254	3755	3003
8	1000	3004	5005	4003
9	1250	3754	6255	5003
10	1500	4504	7505	6003
11	1750	5254	8755	7003
12	2000	6004	10005	8003

Table 6.4. Computational results of Klinz (1989) to compare VERT-PMF and the parametric dual network simplex method.
(a) - number of iterations,
(b) - number of degenerate iterations,
(c) - cpu-time in seconds to calculate the critical intervals,
(d) - total cpu-time in seconds (VAX 11/785).

#	VERT-PMF				Parametric dual NSM			
	(a)	(b)	(c)	(d)	(a)	(b)	(c)	(d)
1	217	113	0.64	3.57	169	65	0.31	0.66
2	430	226	2.04	13.26	333	129	1.02	2.34
3	653	349	5.04	29.81	491	187	1.85	4.84
4	871	467	8.89	52.60	660	256	3.50	8.60
5	2183	1179	57.71	341.17	1666	662	21.37	52.52
6	4247	2243	226.27	1361.26	3369	1365	88.28	213.73
7	6669	3665	533.86	3203.99	5027	2023	200.55	492.15
8	8893	4889	942.97	5676.33	6681	2677	357.29	882.74
9	11107	6103	1469.98	8881.44	8316	3312	559.82	1383.59
10	13322	7319	2111.49	12778.10	9963	3359	806.37	1991.25
11	15531	4596	2868.75	17379.20	11600	4596	1096.81	2710.52
12	17727	9723	3745.74	22720.40	13219	5215	1428.64	3539.97

6.4. Horizontal Algorithm for Parametric Optimal Flows in Generalized Networks

Hamacher & Foulds (1989) developed an approximation approach for PMF calculating $x(t) \in X_{max}(t)$ for all t in [0,T]. The idea is to determine in each iteration a conditional flow augmenting path allowing a flow improvement dx(t) defined on the whole parameter interval. dx(t) is a continuous and piecewise linear function. According to its geometrical interpretation, this is called a horizontal approach.

The worst-case complexity of PMGF investigated in Section 6.2. is the motivation to generalize this idea to parametric generalized networks. For this sake, the vector x of flows is considered to be a vector of linear functions defined in [0,T]. The main contribution of the horizontal approach is that it performs at each iteration an improvement dx(t) which is a piecewise linear function of the parameter. For that reason, a generalization of the notions 'flow augmenting path', 'incremental graph', and '1-generation' respectively 'n-generation' is required. In the case that the calculation is finished before termination of the algorithm, an approximation of the objective value function is at hand.

Let $x(t) = (\ldots,x(i,j;t),\ldots)$ be a given vector of flow functions defined on [0,T]. The sets

$$S^+(i,j;x) := \{t: x(i,j;t) < Cap(i,j;t)\} \text{ and}$$
$$S^-(i,j;x) := \{t: x(i,j;x) > 0\}$$

describe subintervals of [0,T] where a flow augmentation based on x(t) is possible. A *conditional incremental graph* $GC(x) = (V,AC(x))$ is defined according to $AC(x) = AC^+(x) + AC^-(x)$ with

$$AC^+(x) := \{(i,j): (i,j) \in A \ \& \ S^+(i,j;x) \neq \emptyset\} \quad \text{and}$$
$$AC^-(x) := \{(i,j): (j,i) \in A \ \& \ S^-(j,i;x) \neq \emptyset\} .$$

We introduce the abbreviations P^+ and P^- for the sets of arcs $P \cap AC^+(x)$ and $P \cap AC^-(x)$, respectively. By means of

$$S(P^+) := \cap_{(i,j)\in P} \ S^+(i,j;x) \quad \text{and}$$
$$S(P^-) := \cap_{(i,j)\in P} \ S^-(j,i;x)$$

a *conditional flow augmenting path* in GC(x) is defined to be a path P such that $S(P) = S(P^+) \cap S(P^-) \neq \phi$. A closed conditional flow augmenting path L is called a *conditional flow augmenting cycle* with S(L) defined in analogy to S(P). A conditional flow augmenting cycle is called *generating* if $g(P) > 1$ with g(P) defined as in §4. A *conditional n-generation* in GC(x) is a triple (L,k,P), where L is a conditional flow generating cycle containing a vertex k such that there exists a flow augmenting path $P = [k,\ldots,n]$ from k to n with the additional property that $S(L) \cap S(P) \neq \phi$. The *conditional 1-generation* is defined correspondingly. A conditional generating cycle L is called *isolated* with respect to x(t) if there is no vertex k in L such that (L,k,P) is a 1-generation or a n-generation.

A flow x(t) with $x(t) \in \mathbf{X}_g(t)$ for all $t \in [0,T]$ is called *parametric max-min flow*, *parametric n-maximal*, and *parametric n-optimal* if the corresponding properties for non-parametric flows are satisfied for all fixed parameters $t \in [0,T]$.

Theorem 6.4.
The vector x of functions $x(i,j;t)$; $(i,j) \in A$ is a parametric max-min flow iff each conditional flow generating cycle is isolated.

Proof:
For the necessity we assume without loss of generality the existence of a conditional n-generation (L,k,P). Then for all $t \in S(L) \cap S(P)$ the augmentation results in a flow z with $\#(z_n(t)) > \#(x_n(t))$ which, in connection with $\#(z_1(t)) = \#(x_1(t))$ contradicts the definition of parametric max-min flows. The analogous arguments can be applied in the case of a conditional 1-generation. On the other side, if there is no 1-generation and no n-generation for any parameter $t \in [0,T]$, then the application of the non-parametric results of Theorem 4.3. for each fixed parameter t ensures the max-min property of x(t).

■

Theorem 6.5.
Assume a parametric max-min flow x(t) and let P be a conditional augmenting path in GC(x) of maximum multiplicative weight. Then the flow $z(t) = x(t) + dx(t)$ defined by

$$dx(i,j;t) := \begin{cases} 0 & \text{for } (i,j) \notin P \\ (-)\delta(t)\cdot g(P[1,i]) & \text{for } (i,j) \in P^+ \ (\in P^-) \end{cases}$$

with

$\delta(t) := \min \{\delta 1(t), \delta 2(t)\}$ and

$\delta 1(t) := \min \{(Cap(i,j;t) - x(i,j;t))/g(P[1,i]) : (i,j) \in P^+\}$

$\delta 2(t) := \min \{x(i,j;t)/g(P[1,i]) : (i,j) \in P^-\}$
is a parametric max-min flow again.

Proof:
The feasibility of z(t) is a consequence of the corresponding result in Theorem 4.4. Assume that z(t) has not the max-min property. Then there must exist an interval [t1,t2] such that one of the conditions

$y_1(t) \le z_1(t)$ implies $y_n(t) \le z_n(t)$,
$y_n(t) \ge z_n(t)$ implies $y_1(t) \ge z_1(t)$

is violated in [t1,t2] for a feasible flow y(t). Since x(t) is assumed to be a max-min flow, in [t1,t2] it must hold that $dx(t) \ne 0$. However, P is a path of maximum multiplicative weight for all these parameter values $t \in [t1,t2]$, and this contradicts Theorem 4.4.

■

To obtain a necessary and sufficient optimality condition for PMGF, the notion of a cut partitioning is needed. A finite set of 1,n-separating cuts $(X1,X1^*),\ldots,(Xp,Xp^*)$ in connection with a partition I1 + ... + Ip of [0,T] is called a *cut partitioning* (X1,...,Xp;I1,...,Ip).

Theorem 6.6.
x(t) is a solution of PMGF if and only if

(i) there is a cut partitioning (X1,...,Xp;I1,...,Ip) such that for all h = 1,...,p and all $t \in Ih$ it holds that
$(i,j) \in \delta^+(Xh)$ implies $x(i,j;t) = Cap(i,j;t)$,
$(i,j) \in \delta^-(Xh)$ implies $x(i,j;t) = 0$, and

(ii) each conditional flow generating cycle in GC(x) is isolated.

Proof:
For the necessity we assume that x(t) is a solution of PMGF. Then x(t) is parametric n-maximal and has the parametric max-min property. Using Theorem 6.4. confirms the validity of (ii). For (i), a labeling procedure for each of the intervals Ih; h = 1,...,p is performed. Consider the conditional incremental graph GC(x) and a fixed interval Ih; $h \in \{1,\ldots,p\}$. There are three basic steps of the procedure:

```
Initialization: for all i ∈ V do
                begin
                  L(i) := 0
                end
                L(1) := 1
                L1 := {1}
```

```
Iteration:        repeat
                  begin
                    if (i,j) ∈ AC+(x) and L(i) = 1 and L(j) = 0
                       and Ih ⊂ S+(i,j;x) then
                    begin
                      L(j) := 1
                      L1 := L1 + {j}
                    end
                    if (i,j) ∈ AC-(x) and L(i) = 0 and L(j) = 1
                       and Ih ⊂ S-(i,j;x) then
                    begin
                      L(i) := 1
                      L1 := L1 + {i}
                    end
                  end
Termination:      until L(n) = 1 or L1 is maximal.
```

Since x(t) is assumed to be parametric n-maximal, the termination is due to a maximal set L1. For Xh = L1, the conditions (i) of the proposition are satisfied. The labeling procedure is applied with respect to all intervals Ih; h = 1,...,p. Repeating in this way, a cut partitioning (X1,...,Xp;I1,...,Ip) is constructed with the formulated properties.

For the sufficiency, the results of Theorem 4.5. are applied for each of the intervals Ih. Since the union of all these intervals yields in the original interval [0,T], the piecewise linear function x(t) is a solution of PMGF.

■

The last three theorems are the theoretical background of the horizontal algorithm for PMGF. The algorithm consists of two phases. After initialization, conditional n-generations maintaining the feasibility are determined in Phase1. This is done as long as possible. The resulting flow has the parametric max-min property with $\#(x_1(t)) = 0$ for all t according to Theorem 6.5. This process terminates if there is no conditional augmenting path in GC(x) which is equivalent to the existence of a cut partitioning. The optimality is verified by Theorem 6.6.

In Hamacher & Foulds (1989) are described two procedures for calculating conditional flow augmenting paths based on the classical algorithms of Dijkstra (1959) and Bellman (1958). These procedures can be used in Phase2 and can be extended to determine conditional n-generations.

We will illustrate the performance of HORIZONTAL by an example and consider the generalized network of Figure 6.6. The flows and the capacity functions are defined on the interval [0,25]. HORIZONTAL starts with the zero flow $x^0 := 0$. As the underlying graph is acyclic, GC(0) has no flow augmenting cycle. The subsequent sequence of incremental graphs with their flow augmenting paths and the consequences for the capacity functions according to Phase 2 is shown in Figures 6.7. and 6.8. Based on GC(0), the path P = [1,2,4] is found to be the conditional flow augmenting path of maximum multiplicative weight. The function $\delta(t)$ is computed as

$$\delta(t) = \begin{cases} 1 + t & \text{for } t \in [0,2/3] \\ 3/2 + t/4 & \text{for } t \in [2/3,25]. \end{cases}$$

```
procedure HORIZONTAL
begin
x := 0
  repeat ('Phase1')
  begin
    determine a conditional n-generation (L,k,P_L) in GC(x)
    if (i,j) ∈ L then dx(i,j) := char_i,j(L)·g(P_L[k,i])
    if (i,j) ∈ P[k,n] then dx(i,j) := char_i,j(P[k,n])·g(P[k,i])
    δ1(t) := min {(Cap(i,j;t) - x(i,j;t))/g(P_L[k,i])       :(i,j) ∈ L+}
    δ2(t) := min {x(i,j;t)/g(P_L[k,i])                       :(i,j) ∈ L-}
    δ3(t) := min {(Cap(i,j;t)-x(i,j;t)/((g(L)-1)·g(P[k,i]):(i,j) ∈ P+}
    δ4(t) := min {x(i,j;t)/((g(L)-1)·g(P[k,i]))             :(i,j) ∈ P-}
    δ(t)  := min {δ1(t),δ2(t),δ3(t),δ4(t)}
    x(t) := x(t) + δ(t)·dx
  end
  until there is no conditional n-generation in GC(x)
  repeat ('Phase2')
  begin
    P := arg max {g(P*): P* is a conditional flow augmenting path}
    δ1(t) := min {(Cap(i,j;t) - x(i,j;t))/g(P[1,i])           :(i,j) ∈ P+}
    δ2(t) := min {x(i,j;t)/g(P[1,i])                           :(i,j) ∈ P-}
    δ(t)  := min {δ1(t),δ2(t)}
    x(t) := x(t) + δ(t)·char(i,j;P)·g(P[1,i])
  end
end
```

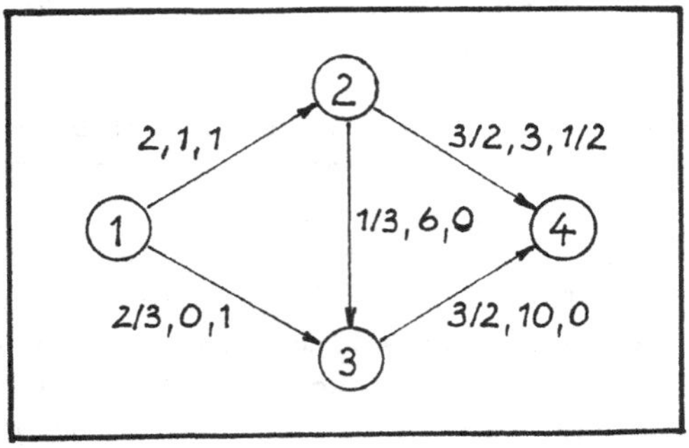

Figure 6.6. Graph G = (V,A). The numbers along the arcs (i,j) denote g(i,j), cap(i,j), and pcap(i,j), respectively.

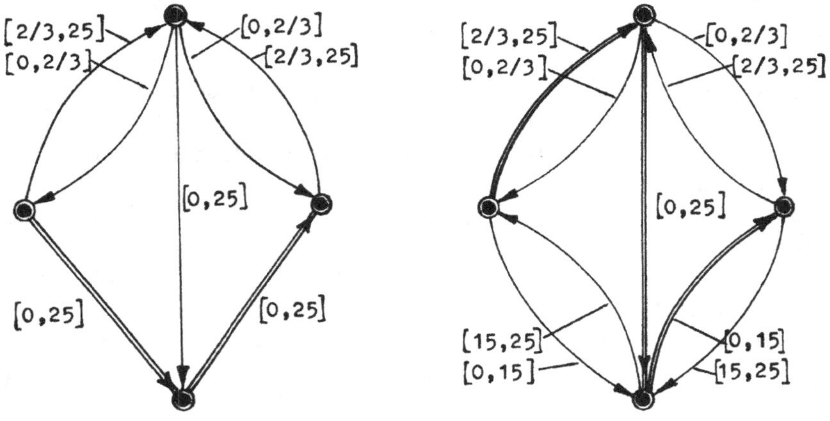

Figure 6.7. Conditional incremental graphs (a) $GC(x^1)$, (b) $GC(x^2)$ with flow augmenting paths of maximum multiplicative weight (double lines) and intervals $S^+(i,j;x)$ and $S^-(i,j;x)$.

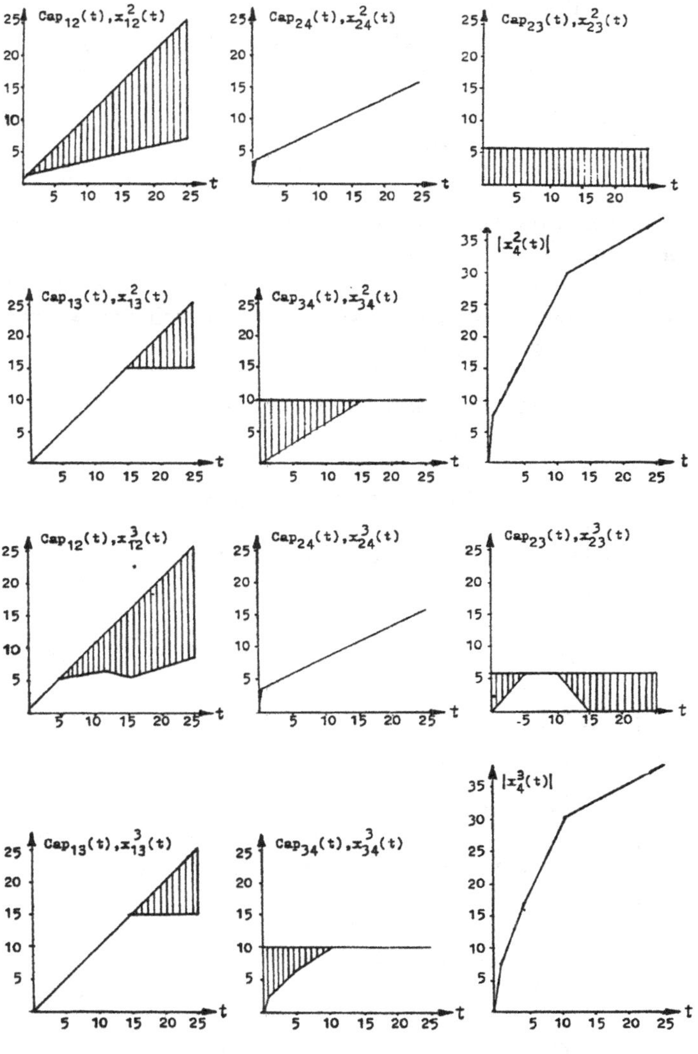

Figure 6.8. Functions Cap(i,j;t) and x(i,;t) for all arcs (i,j) of A. A hatched area indicates free capacity. Additionally, the approximation of the optimal value function $\#(x_4(t))$ is shown.

with bottleneck arcs (1,2) for [0,2/3] and (2,4) for [2/3,25]. At the second iteration, based on $x^1 = \delta(t) \cdot dx$, the path P = [1,3,4] with

$$\delta(t) = \begin{cases} t & \text{for } t \in [0,15] \\ 15 & \text{for } t \in [15,25] \end{cases}$$

is determined. This leads to x^2 and the conditional incremental graph shown on Figure 6.7.(b). In $GC(x^2)$, we obtain the path P = [1,2,3,4]. Figure 6.9. illustrates the process of calculating $\delta(t)$. The function $\delta(t)$ is the inner envelope of the residual capacities res(i,j;t) := (Cap(i,j;t) - x(i,j;t))/g(P[1,i]).

From the numerical calculations, we obtain

$$\delta(t) = \begin{cases} 0 & \text{for } t \in [0,2/3] \\ -1/2 + (3/4)t & \text{for } t \in [2/3,14/3] \\ 3 & \text{for } t \in [14/3,12] \\ 15 - t & \text{for } t \in [12,15] \\ 0 & \text{for } t \in [15,25]. \end{cases}$$

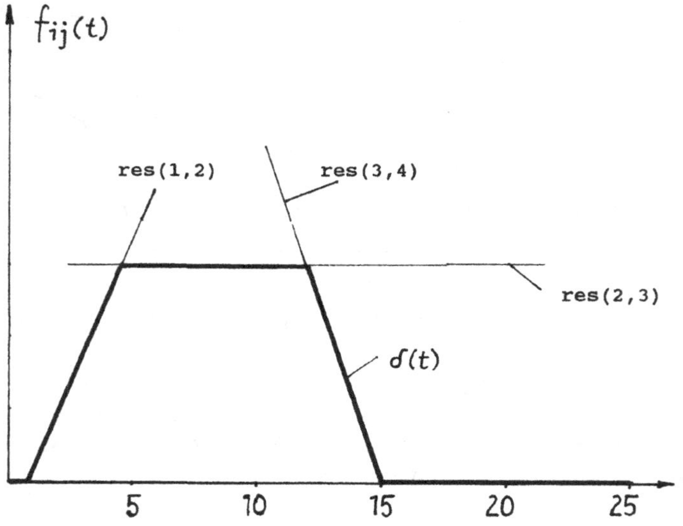

Figure 6.9. Calculation of $\delta(t)$ for x^2 and P = [1,2,3,4].

The three improvements result in

$$\#(x_4(t)) = \begin{cases} 3 + 4t & \text{for } t \in [0,2/3] \\ 4 + (10/4)t & \text{for } t \in [2/3,14/3] \\ 15/2 + (7/4)t & \text{for } t \in [14/3,12] \\ 39/2 + (3/4)t & \text{for } t \in [12,25]. \end{cases}$$

To confirm the parametric n-optimality of this flow, the sufficient conditions of Theorem 6.6. are checked. By means of the intervals I1 = [0,14/3] , I2 = [14/3,12], I3 = [12,25] and the cuts defined by X1 = {1}, X2 = {1,2}, X3 = {1,2,3}, the conditions are satisfied.

6.5. Dual Reoptimization for Parametric Changes in the Minimum-Cost Flow Problem

In general, dual methods seem to be suited for reoptimization in the case that primal feasibility is destroyed and dual feasibility is maintained. There are two approaches of dual algorithms when related to network flow problems. Firstly, one maintains conservation of flow and moves toward primal feasibility by reducing the violation of the lower and upper bounds. Secondly, the feasibility with respect to the capacity constraints is maintained and the fulfillment of the flow conservation rules is gradually improved.

Ali et al. (1989) studied the minimum-cost flow problem

$$\min \ \{ c^T x: \ I(G) \cdot x = b, \ \ 0 \le x \le cap \}$$

and compared dual and primal reoptimization procedures for parameter changes in b, c, and cap. As the base of the comparison, a subset of the benchmark NETGEN problems was used. The dimension of the problems varied between 200 and 1000 nodes and between 2200 and 6320 arcs. Changes in cost and capacities were made on a set of arcs chosen to be either the set of arcs emanating from a node, incident to a node, or a set of arcs chosen across a set of nodes.

The principal steps of the dual network algorithm are given in procedure DUAL. The efficiency of the algorithm strongly depends on the data structure and the effort necessary to select leaving and entering variables at each iteration. We assume that a dual feasible basic solution related to a spanning tree $T = (V, A_T)$ is given.

```
procedure DUAL
begin
  Ainf :=  {(i,j) ∈ AT: x(i,j) > cap(i,j) or x(i,j) < 0}
  repeat
  choose a = (k,l) ∈ Ainf
  determine the fundamental cocycle Ω(T,a) related to a = (k,l)
  for all (i,j) ∈ Ω(T,a) do δ(i,j) := p(j) - p(i) - c(i,j)
  if x(k,l)  >  cap(k,l) then
  begin
    δx := cap(k,l) - x(k,l)
    p1 := - min {- δ(i,j): (i,j) ∈ Ω+(T,a) & x(i,j) = 0}
    p2 := min {δ(i,j): (i,j) ∈ Ω-(T,a) & x(i,j) = cap(i,j)}
    δp := min {p1,p2}
  end
  if x(k,l) < 0 then
  begin
    δx := - x(k,l)
    p1 := min {- δ(i,j): (i,j) ∈ Ω-(T,a) & x(i,j) = 0}
    p2 := min {δ(i,j): (i,j) ∈ Ω+(T,a) & x(i,j) = cap(i,j)}
    δp := min {p1,p2}
  end
C Removal of (k,l) from the current tree results in two trees Tk and Tl
C with k ∈ Tk and l ∈ Tl.
  for all j ∈ Tl do
  begin
    if x(k,l) < 0 then p(j) := p(j) + δp
                  else p(j) := p(j) - δp
  end
C Assume that δp is determined by b = (r,s) ∈ AC.
  determine the basis equivalent path  μ(T,b) such that
    a = (k,l) ∈ μ+(T,b) for x(k,l) < 0 respectively
        (k,l) ∈ μ-(T,b) for x(k,l) > cap(k,l)
  for all (i,j) ∈ μ(T,b) do
  begin
    if (i,j) ∈ μ+(T,b) then x(i,j) := x(i,j) + δx
    if (i,j) ∈ μ-(T,b) then x(i,j) := x(i,j) - δx
  end
  AT := AT - {(k,l)} + {(r,s)}
  update Ainf
  until Ainf = φ
end
```

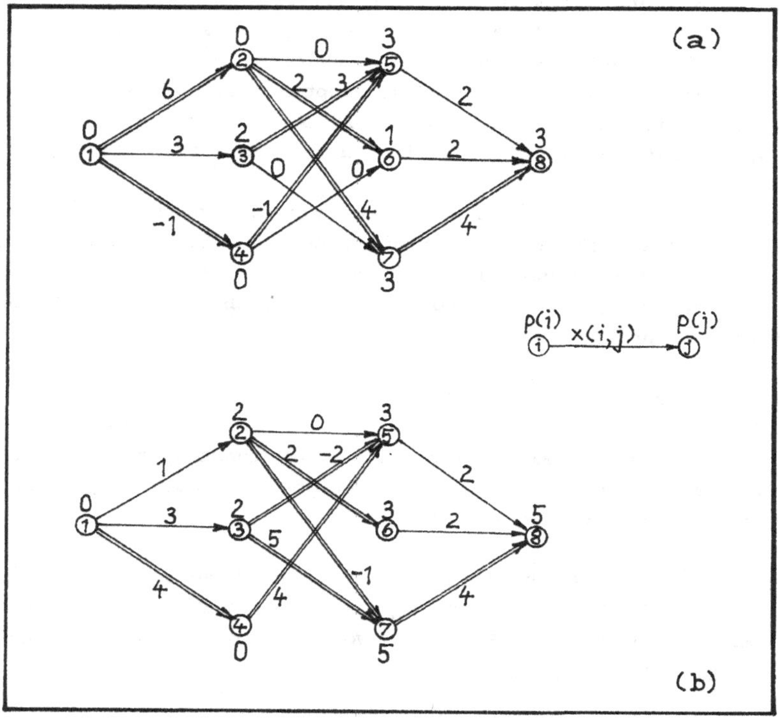

Figure 6.10. (a) Dual feasible basic solution.
(b) Basic solution after basis exchange step.

In Figure 6.10.(a), a dual feasible basic solution related to a spanning tree T for the instance of MCF given in Chapter 3 is presented. The primal feasibility is destroyed by the arcs (1,4) and (4,5) with negative flows and by (1,2) with a flow x(1,2) = 6 > 1 = cap(1,2). If we choose a = (k,l) = (1,2), then the fundamental cocycle Ω(T,a) contains the arcs (1,2),(2,5),(3,7), and (4,6). This results in δp = 2 determined by b = (3,7). The fundamental cycle μ(T,b) is described by the sequence of vertices [3,7,2,1,4,5,3]. The subsequent basic solution is depicted in Figure 6.10.(b).

The computational results obtained by Ali et al. (1989) are promising. For parametric changes in requirements, there was a 93% reduction in the number of pivots and a 53% reduction in time. In the case of parametric changes in capacity, the reduction in the number of pivots was 89% and the time reduction was 32%. Parametric changes in cost lend themselves naturally for primal reoptimization. While the number of dual pivots remains smaller than the number of primal pivots, in some cases the time for dual reoptimization is larger than in primal reoptimization. For sequences of cost changes, there were 75% fewer dual pivots and a 20% reduction in reoptimization time.

The results indicate that dual reoptimization procedures are more efficient than primal reoptimization procedures. The authors finally conclude that dual reoptimization can be applied to advantage in algorithms for large-scale embedded network problems where MCF often occurs as a subproblem.

6.6. Fuzzy Network Flows

The concept of decision making with fuzzy constraints goes back to Bellman & Zadeh (1970). The strong relation to parametric programming was investigated by Chanas (1983).

A *fuzzy set* F on a reference set Y is defined by a membership function μ_F: $Y \longmapsto [0,1]$. $\mu_F(x)$ expresses the membership degree of $y \in Y$ in F. The intersection $F \cap G$ of two fuzzy sets F and G is defined by the membership function $\mu_{F \cap G} = \min \{\mu_F(x), \mu_G(x)\}$.

We consider the fuzzy maximum flow problem in more detail. In this case, the capacity constraints and the objective function are replaced by the corresponding fuzzy formulations. In the first case this means that the precisely determined interval [0,cap(i,j)] for flow x(i,j) goes over to a fuzzy interval. The values $\mu_{ij}(x)$ of the membership function represent the degree of satisfaction of the fuzzy capacity constraint of arc $(i,j) \in A$. We assume linear functions $\mu_{ij}(x)$:

$$\mu_{ij}(x) = \begin{cases} 1 & \text{for } x(i,j) \le cap(i,j) \\ 0 & \text{for } x(i,j) \ge cap(i,j) + p(i,j) \\ 1-t(i,j)/p(i,j) & \text{for } x(i,j) = cap(i,j) + t(i,j), \\ & \quad t(i,j) \in (0,p(i,j)). \end{cases}$$

With respect to the objective function, minimum and maximum desirable values b_0 respectively $b_0 + p_0$ of the total flow #(x) is given by the decision-maker. Again, the (linear) membership function $\mu_0(x)$ represents the degree of satisfaction of the objective function.

$$\mu_0(x) = \begin{cases} 1 & \text{for } \#(x) \geq b_0 + p_0 \\ 0 & \text{for } \#(x) \leq b_0 \\ t_0/p_0 & \text{for } \#(x) = b_0 + t_0,\ t_0 \in (0,p_0). \end{cases}$$

The solution procedure consists in two steps. Firstly, the parametric maximum flow problem is solved with parametric capacities $Cap(i,j;t) = cap(i,j) + t \cdot p(i,j)$ for a parameter interval $t \in [0,1]$. This results in a piece-wise linear, concave function $\#(x^*(t))$. Secondly, the membership function $\mu_{constr}(x^*(t))$ is

$$\mu_{constr}(x^*(t)) = \min\{\mu_{ij}(x^*(t)) : (i,j) \in A\} = 1 - t.$$

The solution $\mu(x^*(t))$ of the fuzzy maximum flow problem is calculated by the application of the minimum operation :

$$\mu(x^*(t)) = \min\{\mu_{constr}(x^*(t)), \mu_0(x^*(t))\}.$$

The application of parametric programming results in a complete spectrum of decisions. The determination of the one with maximum degree of satisfaction is illustrated in Figure 6.11.

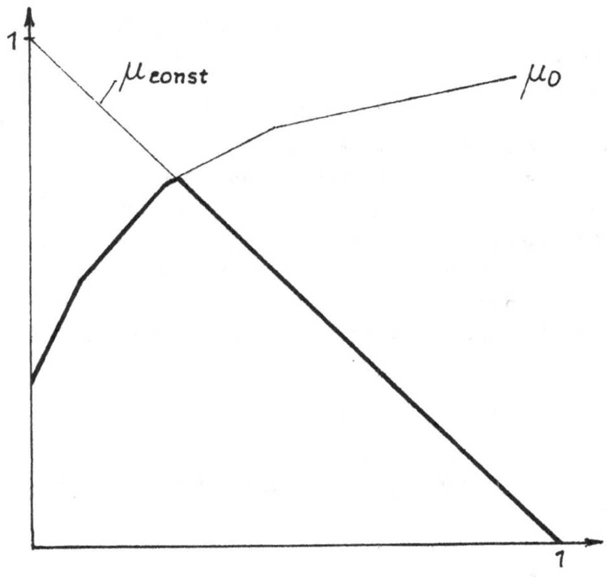

Figure 6.11. Calculation of the maximum degree of satisfaction.

§7. DETECTING NETWORK STRUCTURE

7.1. Embedded Networks, Graph Realization, and Total Unimodularity

The aim in detecting network structure is to convert a matrix of a linear programming problem into a network matrix, i.e., a matrix which contains in each column at most two nonzero elements. If the two nonzero entries are of arbitrary sign and value then we obtain the incidence matrix of a generalized network by column scaling.

The importance of detecting network structure is two-fold. Firstly, there are computational advantages in connection with graph-related formulation of a linear programming problem because in the solution algorithms it is possible to use linked lists, pointers and logical operations in place of the more time consuming arithmetic operations. Numerical investigations in which computer codes for solving network flow problems are compared with the best general-purpose linear programming codes indicate a high superiority of the network codes (up to 200 times faster, see Glover & Klingman 1981). Secondly, the figurative representation of the network structure guarantees a deeper insight into the interrelations between the variables of the problem, which is an essential support for applications.

We distinguish two approaches to find network structure: Embedded networks and graph realization.

An embedded network within a linear program is a subset of constraints and/or variables that can be formulated by conservation rules of pure or generalized network flow problems. Assume B to be the m x n -coefficient matrix of a linear programming problem LP. Define a Boolean matrix H of the same order with $H(i,j) = 1$ if and only if $B(i,j) \neq 0$. The search for embedded network structure which is maximum with respect to the number of rows, columns, and the sum of rows and columns is formulated in the optimization problems EMBED1, EMBED2, and EMBED3, respectively:

EMBED1: $\max \{1^T r: r^T H \leq 2,\ r \in \{0,1\}^m\}$

EMBED2: $\max \{1^T s: H \cdot s \leq 2,\ s \in \{0,1\}^n\}$

EMBED3: $\max \{1^T s + 1^T r: \Sigma_i H(i,j) \cdot r(i) + m(j) \cdot s(j) \leq 2 + m(j)$ for all $j = 1,\dots,n;$ $s \in \{0,1\}^n,\ r \in \{0,1\}^m\}$

with $m(j) := \Sigma_i H(i,j) - 2.$

For graph realization we need the fundamental matrix with respect to a given digraph G and a (spanning) tree T. As defined in Chapter 1, there is a family of fundamental cocycles $\Omega(T,a)$, $a \in T$. From this family we construct the *fundamental matrix* F having rows indexed on T and having columns indexed on A with entries from $\{0,-1,+1\}$ only:

$$(1)\quad F(a,b) := \begin{cases} 1 & \text{if arc } b \text{ is in } \Omega^+(T,a) \\ -1 & \text{if arc } b \text{ is in } \Omega^-(T,a) \\ 0 & \text{if arc } b \text{ is not in } \Omega(T,a). \end{cases}$$

We illustrate the notion of a fundamental matrix in Figure 7.1. The resulting matrix is:

F =

	1	2	5	7	8	3	4	6	9	10
1	1						-1		1	-1
2		1				1	1		-1	1
5			1						1	-1
7				1		-1		1		
8					1	1		-1		1

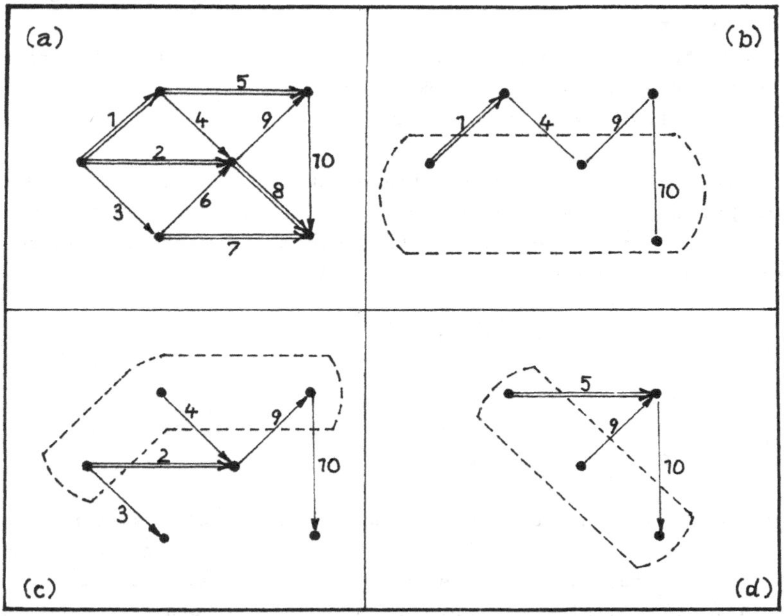

Figure 7.1. (a) Graph G with spanning tree T (double lines).
(b),(c),(d) Cocycles $\Omega(T,a)$ for a = 1,2,5, resp.

A given matrix B is called *graphic* if there is a graph G and a tree T such that $B^* := (E,B)$ where E denotes the identity matrix, is the corresponding fundamental matrix. B is *cographic* if B^T is graphic. *Graph Realization* GR is the problem, given a matrix B, to determine that (E,B) is not the fundamental matrix of a graph, or to construct the graph G and the corresponding T. In general, however, we do not expect to find such a (complete) transformation. In *Partial Graph Realization PG1 and PG2* we ask for a maximum number of rows respectively columns such that the resulting reduced matrix is graphic.

Lemma 7.1. (Ruhe 1988a)
Let F be a fundamental matrix with respect to graph G and tree T. Then

(2) $I(G) = I(T) \cdot F$.

■

As an immediate consequence of this lemma we obtain the important result indicating the equivalence between LP and MCF in the case of graphic matrices B.

Theorem 7.1.
Let B be the coefficient matrix of the linear programming problem

$$\min \{c^T x: B \cdot x \le b\}$$

and assume that B is graphic. Then x^* is a solution of LP if and only if x^* solves the minimum cost flow problem

$$\min \{c^T x: I(G) \cdot x = I(T) \cdot b,\ x \ge 0 \}.$$

■

Graphic matrices and their transpose (the class of cographic matrices) form a large subset of the class of totally unimodular matrices. However, there are totally unimodular matrices, which are neither graphic nor cographic (cf. Bixby 1977):

$$(3) \quad \begin{bmatrix} 1 & -1 & 0 & 0 & -1 \\ -1 & 1 & -1 & 0 & 0 \\ 0 & -1 & 1 & -1 & 0 \\ 0 & 0 & -1 & 1 & -1 \\ -1 & 0 & 0 & -1 & 1 \end{bmatrix}, \quad \begin{bmatrix} 1 & 1 & 1 & 1 & 1 \\ 1 & 1 & 1 & 0 & 0 \\ 1 & 0 & 1 & 1 & 0 \\ 1 & 0 & 0 & 1 & 1 \\ 1 & 1 & 0 & 0 & 1 \end{bmatrix}.$$

Seymour (1980) showed that each totally unimodular matrix arises, in a certain way, from graphic, cographic, and the matrices given in (3). As an important consequence the following theorem can be used as a base of a polynomial-time algorithm for testing total unimodularity:

Theorem 7.2. (Schrijver 1986)
Let B be a totally unimodular matrix. Then at least one of the following cases is valid:

(i) B is graphic or cographic;

(ii) B is one of the matrices given in (3), possibly after permuting rows or columns, or multiplying some rows or columns by -1;

(iii) B has a row or column with at most one nonzero, or B has two linearly dependent rows or columns;

(iv) the rows and columns of B can be permuted such that

$$B = \begin{bmatrix} B1 & B2 \\ & \\ B3 & B4 \end{bmatrix} \quad \text{with} \quad \text{rank}(B1) + \text{rank}(B2) \le 2,$$

where both for B1 and for B4 the number of rows plus the number of columns is at least 4.

■

Clearly, the first three conditions can be tested in polynomial time. Cunningham & Edmonds (1980) proved that there is also a polynomial-time decomposition of B as described in (iv). A recursive application of this decomposition to matrices of increasing size results in the complete algorithm to test for unimodularity. According to Bixby (1982), the implementation of the decomposition test of Cunningham & Edmonds (1980) yields an $O((m + n)^4m)$ algorithm for testing total unimodularity. Therein, m and n are the number of rows respectively columns of the given matrix.

7.2. Complexity Results

Iri (1966) was the first to develop a computational method for the graph realization problem. He presented an $O(m^6)$ algorithm. Further progress was made by Tomizawa (1976) with an $O(m^3)$ algorithm, by Fujishige (1980), and Bixby & Wagner (1985) with almost linear algorithms.

Two matrices A,B are said to be *projectively equivalent* if there are nonsingular matrices C,D such that D is diagonal and $A = C \cdot B \cdot D$. The matrix C performs elementary row operations on B; D is for nonzero column scaling.

Theorem 7.3. (Bixby & Cunningham 1980)
Let B be a matrix with m rows and r nonzero elements. Then it takes $O(m \cdot r)$ steps to decide whether B is projectively equivalent to the vertex-arc incidence matrix I(G) of a graph G. ■

From the motivation given at the beginning it follows that also projectively equivalence between a given matrix and the incidence matrix of a generalized network is of interest. To find a polynomial algorithm for this transformation is an open problem. Chandru, Coullard & Wagner (1985) showed that the problem of determining whether a given linear programming problem can be converted to a generalized network flow problem having no cycles L of unit-weight g(L) = 1 is NP-hard.

All the methods mentioned so far are devoted to answer the question whether a given matrix is completely transformable. However, we do not await this in the most practical cases, and the question is to determine a large subset with the desired properties. The *submatrix problem* for a fixed matrix property π is:

Instance: $m \times n$ - matrix A with elements from $\{0,1\}$;
positive integers k_1, k_2; matrix property π.
Question: Does A contain a $k_1 \times k_2$ submatrix which satisfies π?

A matrix property is *non-trivial* if it holds for an infinite number of matrices and fails for an infinite number of matrices. A matrix property is *hereditary on submatrices* if, for any matrix satisfying the property, all of its submatrices also satisfy the property.

Theorem 7.4. (Bartholdi 1982)
Let π be a property for (0,1)-matrices that is non-trivial, satisfied by all permutation matrices, and is hereditary on submatrices. Then the submatrix problem for π is NP-hard.

■

Among the properties to which Theorem 7.4. can be applied is the property of a matrix to be graphic. Consequently, the above formulations PG1 and PG2 of partial realizability are NP-hard, too. Thus it is unreasonable to require for an exact method for computing a maximum size submatrix with the specified property, say graphicness or cographicness (the same is true for planarity). Truemper (1985) described a very flexible decomposition approach for finding submatrices with certain inherited properties including graphicness, cographicness, planarity, and regularity. If the matrix has the given property, the algorithm will detect this. Otherwise, one may direct the algorithm to find a submatrix that is set-theoretically maximal in the row-sense (i.e., addition of any row to the submatrix destroys the considered property), or that is set-theoretically maximal in the column sense. Truemper's method relies on a series of results about matroid 3-connectivity and matroid decomposition.

7.3. Graph Realization by Means of m-Hierarchies

In Ruhe (1988a) a one-to-one correspondence is developed between the graphicness of a matrix described by its row vectors B(1),...,B(m) and the property of these vectors to be a so-called m-hierarchy as considered by Fedorenko (1975).

We consider a finite set U having four elements 1,-1, 0, and *. A reflexive partial order '»' is defined on U according to

(4) $* \gg 1;\ * \gg -1;\ * \gg 0;\ 1 \gg 0;\ -1 \gg 0;$ and $u \gg u$ for all $u \in U$.

We remark that '»' is also antisymmetric and transitive. The operation '°' (multiplication) is defined as:

(5) $0^\circ u = 0$ for all $u \in U$,
$1^\circ u = u$ for all $u \in U$,
$*^\circ u = *$ for all $u \in \{1,-1,*\}$, and
$(-1)^\circ(-1) = 1$.

Let U_1, U_2 be two vectors of U^n. Then we define

(6) $U_1 \gg U_2$ if $U_1(j) \gg U_2(j)$ for all $j = 1,...,n$.
(7) U_1 «» U_2 if $U_1 \gg U_2$ or $U_2 \gg U_1$ (comparability).
(8) $U_1 \blacksquare U_2$ if for all $j = 1,...,n$:
$U_1(j)^\circ U_2(j) \neq 0$ implies $U_1(j)^\circ U_2(j) = -1$.

Let $H = \{U_1,...,U_m\}$ be a set of vectors of U^n and $\Phi: H \longmapsto V$ a one-to-one mapping between H and the vertices V of a rooted tree $W = (V,A_W)$. H is called a *normal m-hierarchy*, if with $\Phi(U_i) := i\ (\in V)$ the conditions (9) and (10) are fulfilled:

(9) $(i,j) \in A_W$ implies $U_i \gg U_j$;
(10) the set of vertices $V^k := \{i \in V: \#(P_i) = k\}$, where P_i is the unique path connecting the root with i in W, fulfills:
$i,j \in V^k$ implies $U_i \blacksquare U_j$ for all i,j,k.

The above notions are illustrated by an example. Assume $H = \{U_1,...,U_5\}$ with vectors

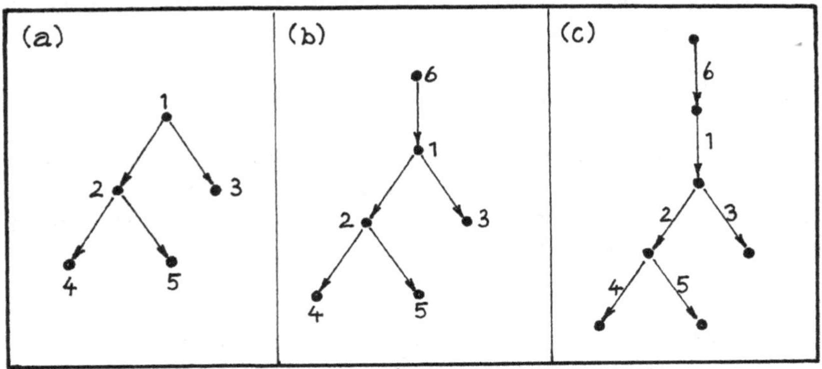

Figure 7.2. Rooted trees (a) W^1, (b) W^2, and (c) Z^2.

$U_1 = (1,*,*,-1)$, $U_2 = (0,*,1,-1)$, $U_3 = (1,0,-1,0)$,
$U_4 = (0,-1,0,1)$, and $U_5 = (0,1,1,0)$.

With the rooted tree of Figure 7.2.(a) and assuming $\Phi(U_i) = i$ for $i = 1,\ldots,5$, the properties (9),(10) of a normal m-hierarchy are satisfied. If we enlarge the set H by one more vector $U_6 = (1,*,0,*)$ then $U_6 \gg U_4$. However, the set H has no maximal element in this case, and consequently, there does not exist an arrangement of the vectors in a rooted tree fulfilling (9) and (10).

Lemma 7.2.
Let $H = \{U_1,\ldots,U_m\}$ be a normal m-hierarchy with rooted tree W.

(i) $H^l(j) := \{U_i \in H: i \in V^l \;\&\; U_i(j) \neq 0\}$ yields in $\#(H^l(j)) \leq 2$.
(ii) For two vectors U_h,U_k of $H^l(j)$ it holds $U_h(j)\cdot U_k(j) = -1$.
(iii) $U_i \in H^l(j)$ and $U_i(j) = *$ implies $U_k(j) = 0$ for all $k \in V^l$.
(iv) $U_h,U_k \in H^l(j)$ and $U_f \in H^{l-1}(j)$ and $U_f(j) = *$ implies (f,h), $(f,k) \in A_W$.
(v) $U_i \in H^l(j)$ and $U_i(j) = *$ implies $U_k(j) = *$ for all k: $k \in P_i$.
(vi) $U_i(j) = 0$ implies $U_k(j) = 0$ for all k: $i \in P_k$.

Proof:

(i) Assume $\#(H^l(j)) > 2$, then (10) cannot be valid.
(ii) With $\#(H^l(j)) = 2$ the proposition is a consequence of (10).
(iii) Otherwise there is a contradiction to (ii).
(iv) (ii) implies $U_h(j)\cdot U_k(j) = -1$. Additionally, $U_g(j) = 0$ for all U_g with $g \in V^{l-1}$ and $U_g \neq U_f$. Then the proposition follows from (9).

(v) For all $k \in P_i$ it is $U_k > U_i$. Consequently $U_k(j) = *$.

(vi) For all $i \in P_k$ it is $U_i > U_k$. Then $U_i(j) = 0$ implies $U_k(j) = 0$. ■

For a given set H of vectors from U^n we define three functions R1, R2, and R3 : $U^n \longmapsto U^n$ which may be applied to fulfill the conditions of a normal m-hierarchy. Let be $u \in H$.

(R1) Replacement of zero elements:
With $J^0(U) := \{j \in J: U(j) = 0\}$ and a subset $J1 \;\; J^0(U)$

define $$U'(j) := \begin{cases} * & \text{for } j \in J1 \\ U(j) & \text{otherwise.} \end{cases}$$

(R2) Multiplication by (-1): $U'(j) := (-1)\cdot U(j)$ for $1 \le j \le n$.

(R3) Replacement of *-elements:
With $J^*(U) := \{j \in J: U(j) = *\}$ and a subset $J2 \;\; J^*(U)$

define $$U'(j) := \begin{cases} 0 & \text{for } j \in J2 \\ U(j) & \text{otherwise.} \end{cases}$$

A given set $H = \{U_1,...,U_m\}$ of vectors from U^n is called an *m-hierarchy* if (R1),(R2),(R3) can be applied in such a way that the resulting vectors form a normal m-hierarchy.

As an example we consider the set of vectors $U_1,...,U_6$ as defined above. As we have seen, this set does not fulfill the conditions of a normal m-hierarchy. However, if we apply R1 with $J1 = \{3\}$ to U_6 resulting in $U_6' = (1,*,*,*)$ then $U_6' \gg U_1$. With the rooted tree of Figure 7.2. (b) the conditions of a normal m-hierarchy are satisfied.

For a given matrix B with elements from {0,1,-1}, we derive vectors $U_i := (B(i,1),...,B(i,n))$ which are the row vectors of B. With respect to all columns j of B we define

$I^+(j) := \{i \in I: B(i,j) = 1\}$ and $I^-(j) := \{i \in I: B(i,j) = -1\}$.

For each rooted tree $W = (V,A_W)$ we introduce a mapping $\Gamma: V \longmapsto A$ such that $Z = (V_Z,A)$ is a rooted tree fulfilling

$(i,j) \in A_W$ implies $\delta^-(\Gamma(i)) = \delta^+(\Gamma(j))$.

The construction of Z^2 is illustrated for the rooted tree W^2 in Figure 7.2.(c).

In the following we present three results indicating the relationship between m-hierarchies and graphic matrices:

Lemma 7.3.
Assume the row vectors $U_1,\ldots,U_m$ of a given matrix form a normal m-hierarchy with rooted tree $W = (V,A_W)$. Then in $Z = (V_Z,\Gamma(V))$ it holds for all $j = 1,\ldots,n$:

(i) $A^+(j) := \{a_i : a_i = \Gamma(i),\ i \in I^+(j)\}$ forms a directed path in Z.

(ii) $A^-(j) := \{a_i : a_i = \Gamma(i),\ i \in I^-(j)\}$ forms a directed path in Z.

(iii) $A(j) := A^+(j) + A^-(j)$ forms a path in Z.

Proof:
Let be $A^+(j) = \{a_1,\ldots,a_q\}$. If there would be two arcs $a_i,a_k \in A^+(j)$ having the same distance (number of arcs) to the root then this implies a contradiction because of $U_i(j) = U_k(j) = 1$. In the case of three arcs a_h,a_i, and a_k such that $a_i,a_k \in A^+(j)$, $a_h \notin A^+(j)$ and $(i,h) \in A_W$, $(h,k) \in A_W$ we obtain a contradiction to (9) from $1 = U_i(j) \gg U_h(j) \gg U_k(j) = 1$ since $U_h(j) \neq 1$. (ii) can be shown analogously.
For (iii) assume that a_1 and a_{1*} are the first (in the direction of the path) arcs of the directed path's of (i) respectively (ii). Then with arcs $(i,1)$ and $(i*,1*)$ of A_W it holds $U_i(j) = *$ and $U_{i*}(j) = *$. From (iii) of Lemma 7.2. it follows that $i = i*$, i.e., $A(j)$ forms a path connecting the two subpaths described in (i) and (ii).

■

Theorem 7.5. (Ruhe 1988a)
Assume the row vectors $H = \{U_1,\ldots,U_m\}$ of a given matrix (E,B) in standard form. If H forms an m-hierarchy with rooted tree $W = (V,A_W)$ then B is graphic with underlying tree Z' where Z' is obtained from Z by reversing all the arcs for which (R2) was applied.

■

In Ruhe (1988a) a hybrid method for detecting a 'large' graphic submatrix by eliminating both rows and columns is developed. Therein, the row-wise approach of the m-hierarchies is combined with a column wise approach using a data structure called PQ-trees. Originally, PQ-trees were introduced by Booth & Lueker (1976) for handling the *Consecutive arrangement problem* :

> Given a finite set F and a collection S of subsets of F. Does there exist a permutation π of the elements of F in which the members of each subset $s \in S$ appear as a consecutive subsequence of π?

For a Pascal-implementation of PQ-tree algorithms compare Young (1977). In connection with graphicness, the consecutiveness of elements is generalized to the consecutiveness of arcs in the realizing tree where the arcs represent the nonzero elements of each column.

7.4. Equivalent Problem Formulations Using Network Flows

Several authors studied the multicommodity transportation problem and asked for the total unimodularity of the constraint matrix. The r-source, s-sink, t-commodity transportation problem is denoted MCTP(r,s,t). Using E for the identity matrix of order $r \cdot s$ and I(G) for the incidence matrix of the bipartite graph G, the constraint matrix A is of the form:

$$A = \begin{bmatrix} I(G) & & & & \\ & I(G) & \ddots & & \\ & & & I(G) & \\ E & E & \cdots & E & E \end{bmatrix} .$$

Rebman (1974) proved that the constraint matrix A^* for a two-commodity transportation problem in which the capacitated edges form a tree with at most one interior node (i.e., all capacitated arcs are incident with a common node) is totally unimodular.

Evans, Jarvis & Duke (1977) proved that a necessary and sufficient condition for the constraint matrix A of MCTP(r,s,t), $t \geq 2$, to be unimodular is that either $r \leq 2$ or $s \leq 2$. Moreover, if possible, the explicit transformation of the problem into an equivalent one commodity flow problem is performed.

Glover & Mulvey (1980) showed how to formulate a zero-one integer programming problem as a mixed integer generalized network and, in some special cases, as a pure network flow problem. The usefulness of these formulations is in providing new relaxations for integer programming that can take advantage of the advances obtained for solving network problems. In more detail, the following integer problem is considered:

$$\textbf{MIP} \quad \min \{c^T x: d \leq Ax \leq b,\ 0 \leq x \leq u,\ x(j) \in \{0,1\} \text{ for all } j \in N_1 \quad N = \{1,2,\ldots,n\}\},$$

where the set $M(j) := \{i \in M: A(i,j) \neq 0\}$ has at most two elements for $j \in N - N_1$. Consequently, if N_1 is empty, MIP is a generalized network flow problem. For $N = N_1$ we have a pure 0-1 integer programming problem. Each variable x(j) is associated with an arc. To create the appropriate network structure, x(j) is subdivided into a collection of arcs which link to each other through a common node. Instead of giving further technical details, the transformation is illustrated by a small example:

$$\min \{ \quad 4x(1) + 5x(2) + 2x(3) - x(4):$$
$$0 \le 2x(1) + 3x(2) + x(3) + 5x(4) \le 7$$
$$0 \le 4x(2) + 2x(3) - x(4) \le 5$$
$$x(1),\ x(2),\ x(3) \in \{0,1\} \quad \}.$$

The corresponding network is depicted in Figure 7.3. The integrality demand on the flow variables x(2,3), x(2,4), x(2,5) in correspondence to the original variables x(1), x(2), x(3), respectively, is indicated by an asterisk. The arcs between vertices from {3,4,5} and {6,7} reflect the inclusion of x(1), x(2), and x(3) (in correspondence to vertices 3,4, and 5) in the first and second constraint (vertices 6 respectively 7). The right hand sides of the two constraints are transformed into the capacity bounds of (6,8) and (7,9). Additionally, a super sink 10 is introduced in the network.

(i,j)	(1,2)	(2,3)	(2,4)	(2,5)	(3,6)	(4,6)	(4,7)
cap(i,j)	3	1	1	1	1	1	1
c(i,j)	0	4	5	2	0	0	0
g(i,j)	1	1	2	2	2	3	4

(i,j)	(5,6)	(5,7)	(7,6)	(6,8)	(7,9)	(8,10)	(9,10)
cap(i,j)	1	1	1	7	5	7	5
c(i,j)	0	0	0	0	0	0	0
g(i,j)	1	2	5	1	1	1	1

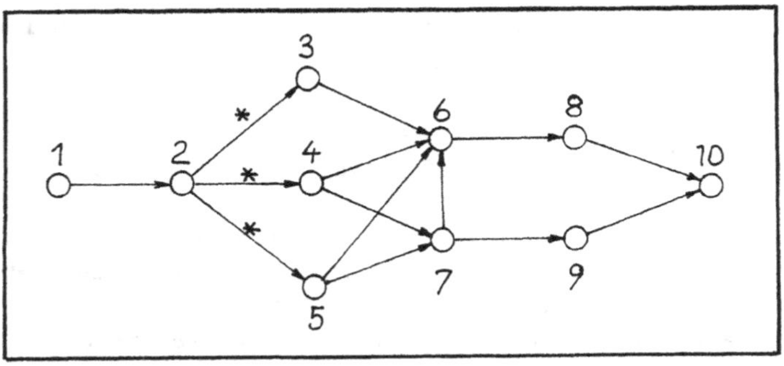

Figure 7.3. Mixed integer generalized network.

7.5. Numerical Investigations to Determine Embedded Networks

We want to determine a subset $I^0 \subset I$ of the set $I = \{1,2,...,m\}$ of row indices such that the correspondingly reduced matrix is the incidence matrix of a generalized flow problem. Just (1987) tested and implemented a heuristic algorithm based on results of Bixby & Fourer (1988). The main idea of the approach is to eliminate at each iteration that row causing the most conflicts to the intended matrix structure. As fundamental operations to achieve that goal, elimination of rows and multiplication of rows by -1 are allowed. The procedure ROWS firstly performs in this way until in each column there are no more than two nonzeros and if two, they have different sign. In the description of the algorithm, cplus(j) and cminus(j) denote the number of positive respectively negative elements in column(j).

To achieve the set-theoretical maximum property, in a subsequent step it is tested whether an already eliminated row with index k can be reincluded. The test is:

```
begin
  for all k ε I - N do
  if (cplus(j) = 0 for all j: B(k,j) > 0 and
     cminus(j) = 0 for all j: B(k,j) < 0)
  or (cplus(j) = 0 for all j: B(k,j) < 0 and
     cminus(j) = 0 for all j: B(k,j) > 0)
  then N := N + {k}
end
```

The heuristic algorithm based there upon was tested on 200 randomly generated examples. In Table 7.1. results for matrices with a fixed number n = 500 of columns and varying in the number m of rows and in the number r of nonzero elements are presented. In each case, the average of five test examples is given.

The design of the algorithm was aimed to produce a compromise between the time necessary to find an embedded network and the size of the network detected. To obtain even better results, different combination of heuristic steps concerning preprocessing, initial extraction and augmentation should be tested.

```
procedure ROWS
begin
  for all i ∈ I do
    p(i) := Σ j:B(i,j)>0 (cplus(j)-1) + Σ j:B(i,j)<0 (cminus(j)-1)
  I^0 := I
  repeat
  begin
    k := arg max {p(i): i ∈ I^0}
    r(k) := Σ j:B(k,j)>0 cminus(j) + Σ j:B(k,j)<0 cplus(j)
    if r(k) < p(k) then
                    begin
                      for all j ∈ J: B(k,j) > 0 do
                      begin
                        cplus(j) := cplus(j) - 1
                        cminus(j) := cminus(j) + 1
                      end
                      for all j ∈ J: B(k,j) < 0 do
                      begin
                        cplus(j) := cplus(j) + 1
                        cminus(j) := cminus(j) - 1
                      end
                    end
                    else
                    begin
                      I^0 := I^0 - {k}
                      for all j ∈ J: B(k,j) > 0 do
                      begin
                        cplus(j) := cplus(j) - 1
                        cminus(j) := cminus(j) - 1
                      end
                    end
    for all i ∈ I^0 do update p(i)
  end
  until p(i) ≤ 0 for all i ∈ I^0
end
```

Table 7.1. Number of embedded network rows.

m\r	200	300	400	500	600	700	800	900	1000
100	95	84	73	64	54	46	39	36	29
200	196	182	171	154	138	124	110	99	86
300	297	285	267	250	232	210	193	171	158
400	396	385	368	352	325	308	277	257	240
500	496	489	468	451	423	396	368	341	312

In addition to the above investigations the question was considered to eliminate the most conflicting columns such that this results in a larger embedded network when the sum of the number of rows and the number of columns is considered. For this sake, two classes of matrices of dimension (a) 200x500 and (b) 300x500 with 600 respectively 700 nonzero elements were taken into account. In dependence of the number of eliminated columns, both the number of rows and the sum of the number of rows and the number of columns of the embedded network are presented in Table 7.2.

Table 7. 2. Effect of column elimination.

# el. col.	(a) # rows	(200,500,600) # rows + # col.	(b) #rows	(300,500,700) #rows + #col.
0	140	640	216	716
10	151	641	227	717
20	161	641	240	720
30	168	638	250	720
40	176	636	257	717
50	184	634	264	714
60	190	630	271	711
70	193	623	280	710
80	194	614	289	709
90	194	604	291	701
100	196	506	292	692

§8. SOLUTION OF NETWORK FLOW PROBLEMS WITH ADDITIONAL CONSTRAINTS

8.1. Introduction

The achieved success in solving pure or generalized network flow problems has caused the question to investigate more general linear programs with embedded network structure. The objective is to transform as much as possible of the efficiency of the basic flow routines to the solution algorithm for the more general case. Due to Glover & Klingman (1981), most large-scale LP-problems involving production scheduling, physical distribution, facility location, or personal assignment contain a large embedded network component, sometimes consisting of several smaller embedded networks. The general linear constraints arise, for example, from economies of scale, capacity restrictions on modes of transportation, limitations on shared resources, or from combining the outputs of subdivisions to meet overall demands.

To give a formal problem statement, we consider the problem

FAC min $\{c^T x: A \cdot x = b,\ 0 \le x \le u\}$ with matrix A given by

$$A = \left[\begin{array}{c|c} I(G) & 0 \\ \hline S & P \end{array}\right] \quad \begin{array}{c} (n \text{ rows}) \\ \hline (q \text{ rows}) \end{array} .$$

The important part of A is the vertex-arc incidence matrix I(G) of graph G. For q = 0 respectively n = 0, FAC goes over into the minimum-cost flow problem and the general linear programming problem. Algorithms for the solution of pure network problems with side constraints have been considered by Klingman & Russell (1975), Chen & Saigal (1977) and Barr et al. (1986). An even more general case allowing the incidence matrix $I_g(G)$ of a generalized flow problem instead of I(G) was considered by McBride (1985).

In the case q = 1 of only one additional constraint, special results and algorithms were developed. Klingman (1977) investigated the question to transform an additional linear constraint into a bounded sum of flow variables associated with arcs entering or leaving a single vertex. A procedure is described to determine if there exists a linear combination of the flow conservation constraints which, when subtracted from the extra constraint, yields a bounded sum of variables associated with arcs entering or leaving a single vertex. Such bounded sums were shown to have equivalent formulations as node conservation constraints.

Belling-Seib et al. (1988) compared three solution methods for FAC in the case of only one additional constraint. The first method is a specialized simplex algorithm which exploits the underlying network structure. The second one is a straight forward dual method which successively reduces the unfeasibility of the side constraint. Finally, a Lagrangean approach is tested which uses a relaxation of the side constraint. Computational experience indicates that the specialized primal simplex algorithm is superior to the other approaches for all but very small problems.

If all data of FAC with $q = 1$ are integer and the number n of vertices in the network is an upper bound for the cost coefficients, Brucker (1985) developed a polynomial algorithm solving $O(|\log n|)$ pure circulation problems LMCF in which two objectives are lexicographically to be minimized.

Koene (1984) introduced so-called *processing networks* as minimum-cost flow problems in which proportional flow restrictions are permitted on the arcs entering or leaving a node. A refining node is a node with one incoming arc and at least two outcoming arcs. The flow on each outgoing arc is required to be a given fraction of the flow entering the vertex. The related constraints are:

$x(j,k(r)) = \alpha(j,k(r))\cdot x(i,j)$ for $r = 1,\dots,R$ with
$0 < \alpha(j,k(r)) < 1$ for all r and
$\Sigma_r\ \alpha(j,k(r)) = 1.$

A blending node is a node with at least two incoming arcs and only one outcoming arc. In this case, the flow on each incoming arc is required to be a given fraction of the total flow having the node. The related constraints are:

$x(k(s),j) = \alpha(k(s),j)\cdot x(i,j)$ for $s = 1,\dots,S$ with
$0 < \alpha(k(s),j) < 1$ for all s and
$\Sigma_s\ \alpha(k(s),j) = 1.$

Chang, Chen & Engquist (1989) developed a primal simplex variant for processing networks maintaining a working basis of variable dimension. In testing against MPSX/370 on a class of randomly generated problems, a Fortran implementation of this algorithm was found to be an order-of-magnitude faster.

8.2. A Primal Partitioning Algorithm

The most popular algorithm for the general problem FAC is the primal partitioning procedure. The main idea is to apply the simplex method where the basis and its inverse are maintained in partitioned form. This allows to handle a part of the basis in correspondence to a spanning tree. Barr et al. (1986) combine this approach with an LU factorization of the basis inverse as known from modern linear programming systems. Kennington & Whisman (1988) presented a Fortran code called NETSIDE for solving FAC which is mainly based on the results of Barr et al. (1986). In the following we give a description of the main steps of the primal simplex method as developed by Barr et al.

For the basic description of the primal simplex algorithm we assume that a feasible solution is calculated by introducing dummy variables and by performing a 'phase 1'. Given a feasible solution, the matrix A and vectors c, x, and u may be partitioned into basic and nonbasic components. We will use the notation $A = (B|N)$, $c = (c^B|c^N)$, $x = (x^B|x^N)$, and $u = (u^B|u^N)$. Since the rank of I(G) is one less than n, a so-called root arc is added. Without loss of generality it is assumed that the underlying graph G is connected, i.e., there is a path between every pair of vertices. Additionally, the matrix $[S|P]$ must have full row rank, otherwise artificial variables are incorporated. A(i) and x(i) denotes the i-th column of the matrix A and the i-th component of vector x, respectively. 1_p denotes a vector with a 1 in the p-th position and zeros elsewhere. Now, formulation FAC can be written as

$$
\begin{aligned}
(1)\ \min \{\ & c_1^T \cdot x_1 + c_2^T \cdot x_2 : \\
& I(G) \cdot x_1 \qquad\qquad + 1_p \cdot a = b^1 \qquad \text{for } 1 \le p \le n, \\
& \quad S \cdot x_1 + \ P \cdot x_2 \qquad\quad = b^2 \\
& \quad\quad x_1 \ge 0, \quad x_2 \ge 0, \\
& \qquad\qquad\qquad\qquad 0 \le a \le 0 \\
& 0 \le x_1 \le u_1, \\
& \qquad\quad 0 \le x_2 \le u_2 \qquad\qquad\qquad\qquad \}.
\end{aligned}
$$

It is well known that every basis B of A may be placed in the form

$$(2)\quad B = \left[\begin{array}{c|c} T & C \\ \hline D & F \end{array}\right],$$

where T is a submatrix of $(M|1_p)$ and $\det(T) \neq 0$. T corresponds to a rooted spanning tree. The basis inverse B^{-1} is of the form

$$(3)\quad B^{-1} = \left[\begin{array}{c|c} T^{-1} + T^{-1}\cdot C\cdot Q^{-1}\cdot D\cdot T^{-1} & - T^{-1}\cdot C\cdot Q^{-1} \\ \hline - Q^{-1}\cdot D\cdot T^{-1} & Q^{-1} \end{array}\right],$$

where $Q := F - D\cdot T^{-1}\cdot C$. Q is referred to as the *working basis*. Barr et al.(1986) describe an algorithm to maintain the inverse of this working basis as an LU factorization. Their specialized code exploits not only the network structure but also the sparsity characteristics of the working basis.

To calculate the dual multipliers $\pi = c^B\cdot B^{-1}$ we assume a partitioning in accordance with (2) and (3). Using the basis inverse (3) results in

$$\begin{aligned}\pi &= (\pi_1|\pi_2)\\ &= ((c_1{}^B + c_1{}^B\cdot T^{-1}\cdot C\cdot Q^{-1}\cdot D - c_2B\cdot Q^{-1}\cdot D)T^{-1} \mid (c_2{}^B - c_1{}^B\cdot T^{-1}\cdot C)Q^{-1}).\end{aligned}$$

Concerning the calculation of the updated column, two cases are distinguished. Firstly, suppose that the entering column is of the form

$$(4)\quad N(k) = \left[\begin{array}{c} 0 \\ \hline P(j) \end{array}\right].$$

Then the updated column $y := B^{-1}\cdot N(k)$ has the form

$$y = \left[\begin{array}{c} y^1 \\ \hline y^2 \end{array}\right] = \left[\begin{array}{c} - T^{-1}\cdot C\cdot Q^{-1}\cdot P(j) \\ \hline Q^{-1}\cdot P(j) \end{array}\right].$$

Secondly, consider the case that N(k) is of the form

$$(5)\quad N(k) = \begin{bmatrix} M(j) \\ \hline S(j) \end{bmatrix}$$, where M(j) denotes the j-th column of I(G).

Then the updated column $y := B^{-1} \cdot N(k)$ has the form

$$y = \begin{bmatrix} y^1 \\ \hline y^2 \end{bmatrix} = \begin{bmatrix} T^{-1}[M(j) + C \cdot Q^{-1} \cdot D \cdot T^{-1} \cdot M(j) - C \cdot Q^{-1} \cdot S(j)] \\ \hline Q^{-1} \cdot [S(j) - D \cdot T^{-1} \cdot M(j)] \end{bmatrix} .$$

In the subsequent formal description of the algorithm called BFK we use

$$\sigma(j) := \begin{cases} 1 & \text{if } y(j) > 0 \\ -1 & \text{if } y(j) < 0 \\ 0 & \text{otherwise} \end{cases} .$$

```
procedure BFK
begin
  repeat
  begin
    Γ := c1B·T-1
    α := c2B - Γ·C
    π2 := α·Q-1
    α := π2·D
    π1 := Γ - α·T-1
    θ1 := {i: xN(i) = 0 & π·N(i) > cN(i)}
    θ2 := {i: xN(i) = uN(i) & π·N(i) < cN(i)}
    choose k ∈ θ1 + θ2
    if k ∈ θ1 then sgn := 1 else sgn := -1
    if N(k) is of the form (4) then
                                    begin
                                      y2 := Q-1·P(j)
                                      Γ := C·y2
                                      y1 := -T-1·Γ
                                    end
    if N(k) is of the form (5) then
                                    begin
                                      Γ1 := T-1·M(j)
                                      Γ2 := S(j) - D·Γ1
                                      y2 := Q-1·Γ2
                                      Γ2 := C·y2
                                      y1 := Γ1 - T-1·Γ2
                                    end
    d1 := min {min {xB(j)/abs(y(j)): σ(j) = sgn},∞}
    d2 := min {min {[uB(j) - xB(j)]/abs(y(j)): σ(j) = sgn},∞}
    delta := min {d1,d2,uN(k)}
    if delta = ∞ then the problem is unbounded
    xN(k) := xN(k) + delta·sgn
    xB := xB - delta·sgn·y1
    θ3 := {j: xB(j) = 0 & σ(j) = sgn}
    θ4 := {j: xB(j) = uB(j) & - σ(j) = sgn}
    choose l ∈ θ3 + θ4
    B(l) := N(k)
    update B-1
  end
  until θ1 + θ2 = φ
end
```

8.3. Solution of a Class of Interval Scheduling Problems

In this chapter we investigate a special scheduling problem having fixed start and finish times. The problem is transformed into a network flow problem with additional constraints and integrality demands.

Between the time of arrival and the time of departure from an airport, an airplane must be inspected before being allowed to start again. By neglecting the stochastic elements, such an inspection can be seen as a job having a fixed start and finish time and being of a certain job class. The start time and the finish time of a job might coincide with the time of arrival and the time of departure of the aircraft but this is not necessary: a list of maintenance norms is available which can be used for calculating the start and finish time of each job.

The inspections should be carried out by ground engineers. An engineer is allowed to carry out the inspection of a specific aircraft only if he has a license for the corresponding aircraft type. From the point of view of operational management of the engineers it would be optimal if each engineer would have a licence for each aircraft type. In that case the engineers could be considered as being qualitatively identical and consequently, each job could be assigned to each of them. However, this situation cannot be realized in practice, as a governmental rule states that each engineer is allowed to have two licences at most.

Within this context, both tactical and operational questions should be answered. Two such questions are:

(i) How many engineers should be available for carrying out all the jobs and what combinations of licences should each of them have?

(ii) How to assign the jobs to the engineers,if both the number of available engineers and the combination of licences of each engineer is given?

The latter problem is studied in Kroon & Ruhe (1989). The operational feasibility problem, asking whether there exists a feasible schedule for all jobs (or not) is called *Class Scheduling* or CS for short. The operational optimization problem, asking for a subset of jobs of maximum total value in a feasible schedule is called *Maximum*

Class Scheduling or MCS. In order to follow the standard terminology on scheduling we will use "machine" and "job" instead of the "mechanic" respectively "inspection".

Suppose there is a set $J = \{1,...,n\}$ of n jobs that have to be carried out within a given time interval $[0,T]$. Job $j \in J$ requires continuous processing in the interval $[s_j,f_j]$. The jobs can be divided into p job classes. Each job j falls into exactly one job class u_j and has a value w_j expressing its importance. Therefore, each job $j \in J$ can be represented by a quadruple of integers (s_j,f_j,u_j,w_j).

The jobs must be carried out in a non-preemptive way by a number of machines, each of which is available in the time interval $[0,T]$. The machines can be split up into a number of machine classes. The number of machine classes is denoted by q.

The p x q - zero-one matrix L with rows and columns in correspondence to job respectively machine classes specifies in each entry L(k,l) whether it is allowed to assign a job of class k to a machine of class l. The assignment is possible if and only if L(k,l) = 1. We give three examples of matrices L:

$$L_0 = (\ 1\ )\ , \quad L_1 = \begin{bmatrix} 1 & 1 \\ 1 & 0 \\ 0 & 1 \end{bmatrix}\ , \quad L_2 = \begin{bmatrix} 0 & 1 & 1 \\ 1 & 0 & 1 \\ 1 & 1 & 0 \end{bmatrix}.$$

In the case of L_0, all jobs are of the same class and can be carried out by all the machines which are of the same class. L_2 represents a situation with three job and three machine classes where each machine class has exactly two licenses.

Class Scheduling CS(L)

Instance: Set of jobs J and for each $j \in J$ a triple (s_j,f_j,u_j). Integers $m_1,...,m_q$, representing the number of available machines in the q machine classes.

Question: Does there exist a feasible, non-preemptive schedule for all the jobs?

In the problem CS(L) the only question is whether there exists a feasible schedule or not. If there exists a feasible schedule, then it is clear, that the maximum number of jobs that can be scheduled is equal to the total number of jobs. However, if a feasible schedule does not exist, then one might ask for the maximum number of jobs or, more generally, for a subset of jobs of maximum total value. Then the problem MCS(L) arises:

Maximum Class Scheduling MCS(L)

Instance: Set of jobs J and for each $j \in J$ a quadruple (s_j, f_j, u_j, w_j). Integers $m_1, \ldots, m_q$, representing the number of available machines in the q machine classes.

Question: Determine a feasible, non-preemptive schedule for a subset $J^0 \subset J$ of jobs such that the total value $w(J^0) := \Sigma_{j \in J^0} w_j$ is maximum.

Note that the matrix L is not part of the instance in both CS(L) and MCS(L).

The computational complexity of both class and maximum class scheduling strongly depends on the structure and dimension of the underlying matrix L. In the case $L = L_0$ the problem CS(L) is equivalent to the well known *Fixed Job Scheduling Problem* FSP which has been studied by many authors and is solvable in polynomial time (see Dantzig & Fulkerson 1954 or Gupta, Lee & Leung 1979). In Lemma 8.1., we combine the equivalence between $CS(L_0)$ and FSP with a result of Dantzig & Fulkerson (1954). The *maximum job overlap* is defined to be the maximum number of jobs that must be carried out simultaneously.

Lemma 8.1.
For an instance of $CS(L_0)$ a feasible schedule for all the jobs exists if and only if the maximum job overlap is less than or equal to the number of available machines.

■

It is well known (see Gupta, Lee & Leung 1979) that the maximum job overlap can be calculated in $O(n \cdot \log n)$ time using appropriate data structures. Arkin & Silverberg (1987) show that $MCS(L_0)$ can be solved by finding a minimum-cost flow in a graph having O(n) vertices and O(n) arcs. Consequently, also $MCS(L_0)$ can be solved in strongly polynomial time (compare Chapter 3).

Kolen & Kroon (1989) investigate the complexity of more general cases of L and develop a complete classification. They use the concept of (ir)reducibility of a matrix which allows to study only the essential cases. A matrix L is called *reducible* if at least one of the following conditions is satisfied:

(i) L contains a row or column having zero's only.
(ii) L contains two identical rows or two identical columns.
(iii) By applying row or column permutations, L can be transformed into a block-diagonal matrix:

$$L = \begin{bmatrix} L1 & 0 \\ & \\ 0 & L2 \end{bmatrix}.$$

If none of these conditions is satisfied, L is called irreducible.

Theorem 8.1 (Kolen & Kroon 1989)
Let L be an irreducible zero-one matrix.
(i) CS(L) is NP-complete if and only if $q \geq 3$.
(ii) MCS(L) is NP-hard if and only if $q \geq 2$.

■

For matrices L corresponding to practical situations as described in the first section, the problems CS(L) and MCS(L) are NP-complete respectively NP-hard almost always. For the matrices defined above it follows that CS(L_2) is NP-complete and MCS(L_1) is NP-hard.

A generalization of the problem MCS(L) is given by Arkin & Silverberg (1987). They assume, that all the jobs have a fixed start and finish time and a weight. They omit the assumption that the number of job classes and the number of machine classes are fixed beforehand: In their setting for each individual job a subset of individual machines is given, and it is assumed, that only machines in this subset may be used for carrying out the job. Arkin and Silverberg show that the problem to maximize the total weight of performable jobs is NP-complete. They also present an algorithm of time complexity $O(n^{b+1})$, where b denotes the total number of available machines. The existence of this algorithm proves that the problem can be solved in polynomial time if the number of machines is fixed beforehand. This result can also be applied to the problems CS(L) and MCS(L).

In the following, we give an alternative formulation of MCS(L) which allows a better algorithmic treatment. The final mathematical model is an integer minimum-cost flow problem with as many additional constraints as there are jobs which can be performed by more than one machine class. The general formulation of the integer minimum-cost flow problem with additional constraints is:

$$\textbf{IFAC} \quad \min \{ c^T x : \; I(G)\cdot x = b, \; 0 \le x \le cap, \; D\cdot x \le d, \; x \in \mathbb{Z}^m \},$$

where

- $I(G)$ is the vertex-arc incidence matrix of digraph G,
- c is the cost vector defined on A,
- b is the demand-supply vector defined on V with $b(i) < 0$ indicating a supply and $b(i) > 0$ indicating a demand vertex for all $i \in V$,
- cap is the vector of upper capacity bounds defined on A,
- D,d is the matrix respectively the vector of right sides of the additional constraints.

A transformation from MCS(L) to a problem called IFAC(L) will be described. The underlying graph $G = (V,A)$ is composed of q subgraphs G^k; $k = 1,...,q$ in correspondence to the q machine classes. For each arc $a \in A$ we denote by δ^+a and δ^-a the head respectively the tail of a. Each subgraph G^k has one source 1^k and one sink n^k. With each job $j \in J^k$ performable in the k-th machine class three arcs

$$(1^k,\delta^-a_j{}^k),(\delta^-a_j{}^k,\delta^+a_j{}^k),(\delta^+a_j{}^k,n^k)$$

are associated. Additionally, arcs $(\delta^+a_i,\delta^-a_j)$ are introduced whenever $f_i < s_j$ for $i,j \in J^k$. The composition of the subgraphs G^k is done by introducing a super source 1 and a super sink n connected with all sources 1^k respectively all sinks n^k resulting in arcs $(1,1^k),(n^k,n)$ for $k = 1,...,q$. The vectors b, cap, and c are defined according to (6) - (9) with

$$\beta = \sum_{k=1}^{q} m_k :$$

$$(6) \qquad b(j) = \begin{cases} \beta & \text{for } j = n \\ -\beta & \text{for } j = 1 \\ 0 & \text{otherwise .} \end{cases}$$

$$(7) \quad cap(u,v) = \begin{cases} m_k & \text{for } (u,v) = (1,1^k),(n^k,n) \\ 1 & \text{otherwise .} \end{cases}$$

$$(8) \qquad c(u,v) = \begin{cases} -w_j & \text{for } (u,v) = (\delta^- a_j{}^k, \delta^+ a_j{}^k) \\ 0 & \text{otherwise .} \end{cases}$$

With J^0 to be the set of jobs which can be processed by more than one machine class, D is an card(J^0) x card(A) - matrix with elements from {0,1} only. d is a vector of dimension card(J^0). The elements D(j,k) of the matrix D are:

$$(9) \qquad D(j,h) = \begin{cases} 1 & \text{if job } j \in J^0 \text{ is represented by arc } a_h \in A \\ 0 & \text{otherwise .} \end{cases}$$

In a row of D there are as many nonzero entries as we have possibilities that the job in correspondence to that row can be carried out by machines of different classes. Replacing (1) - (4) in IFAC in connection with the given matrix L leads to the problem called IFAC(L).

Lemma 8.2

Each optimal solution x of IFAC(L) implies an optimal solution of MCS(L) where job $j \in J$ is processed by a machine of class k if and only if $x(\delta^- a_j{}^k, \delta^+ a_j{}^k) = 1$.

Proof:

(i) Feasibility: Any solution of IFAC(L) contains only 0,1-components for all arcs not incident to both 1 and n. Each job is processed by at most one machine. From the topological structure of G it follows that the jobs of each machine do not overlap.

(ii) Optimality: Assume the contrary. Then from a schedule containing more jobs than the one derived from x we easily obtain an integer solution x' with $c^T x' < c^T x$.

■

The application of the transformation procedure is illustrated by an example. There are ten jobs of three job classes with parameters as described in Table 8.1 and a matrix $L = L_1$ as given above. The jobs are assumed to be of equal importance, i.e., $w_j = 1$ for all $j \in J$. We propose that $m_1 = 2$ and $m_2 = 2$ machines are available. From the application of the transformation we obtain the graph shown in Figure 8.1.

Table 8.1

job	j	1	2	3	4	5	6	7	8	9	10
start time	s_j	5	10	2	9	2	2	0	0	7	1
finish time	f_j	9	12	6	15	9	5	9	6	12	8
job class	u_j	2	1	3	2	3	2	2	3	1	1

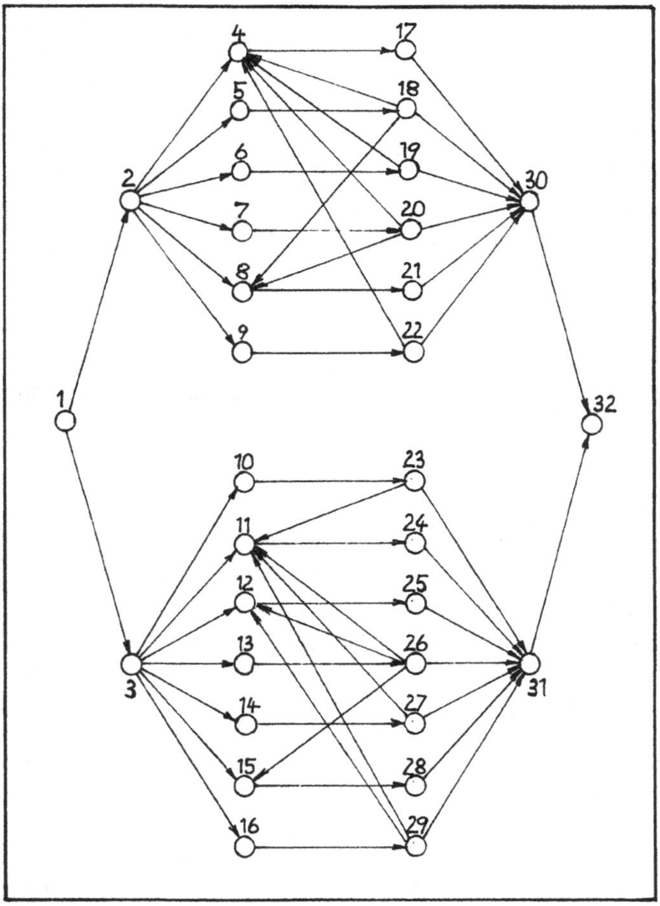

Figure 8.1. Graph G = (V,A) in correspondence to the example given in Table 8.1.

For the minimum-cost flow problem with additional constraints, the total unimodularity of the problem without additional constraints is no more valid. Even worse, the integer problem is known to be NP-complete since it contains the integer knapsack problem as a special case.

Lagrangean relaxation and surrogate constraints haven been proven to be as techniques suited to solve special structured integer optimization problems for which no efficient computational scheme is known or can be awaited. Let $\mathbf{X}$ be the flow polyhedron defined as in Chapter 3. We will denote IFAC as problem P for short and write it in the form:

$$z(P) = \min \{c^T x: x \in \mathbf{X}, x \in \mathbf{Z}^m, Dx \le d\}.$$

A surrogate relaxation of problem P associated with a given multiplier vector $\mu \ge 0$ is

$$z(P_\mu) = \min \{c^T x: x \in \mathbf{X}, x \in \mathbf{Z}^m, \mu^T Dx \le \mu^T d\}.$$

$z(P_\mu)$ is obviously a lower bound on $z(P)$ for all $\mu \ge 0$. The best such bound derives from the surrogate dual

$$z(D_S) = \max \{z(P_\mu): \mu \ge 0\}.$$

The more widely known Lagrangean relaxation with a given vector τ is

$$z(P^\tau) = \min \{c^T x + \tau^T (Dx - d): x \in \mathbf{X}, x \in \mathbf{Z}^m\}.$$

Again, $z(P^\tau)$ is a lower bound on $z(P)$ for each $\tau \ge 0$. Its corresponding Lagrangean dual is

$$z(D^L) = \max \{z(P^\tau): \tau \ge 0\}.$$

Karwan & Rardin (1979) have compared the bound obtained from surrogate constraints and Lagrangean relaxation in integer programming. They have shown that a gap $z(P_S) - z(P^L) > 0$ is to be expected in all but some very special cases. Based on this result and some first empirical tests in solving IFAC(L) we decided to apply the surrogate constraint approach.

Lemma 8.3.
Suppose that $\mu \geq 0$ and that x^* is a solution of P_μ. If $Dx^* \leq d$ then x^* is an optimal solution.

Proof:
For each $x \in \mathbf{X}$: $Dx \leq d$ it follows that also $\mu^T Dx \leq \mu^T d$. Since the objective functions of P and its surrogate relaxation P_μ are the same, the optimality of x^* for the relaxed problem in conjunction with feasibility to the original problem implies optimality with respect to P.

■

There are two main problems in solving the surrogate dual. Firstly we need an effective solution method for the surrogate relaxation which is a minimum-cost flow problem with one additional constraint and integrality demand. The reduction in the number of additional constraints has not changed the NP-completeness of this flow problem since integer knapsack is a special case as before. Secondly, a satisfactory method for the calculation of the consecutive dual multipliers μ must be found.

The proposed algorithm applies parametric programming to obtain two sequences of solutions which lead to lower respectively upper bounds of z(P). For that reason we assume a surrogate multiplier $\mu \geq 0$ and consider the parametric minimum-cost flow problem $S(\mu,t)$ with

(10) $z(S(\mu,t)) = \min \{c^T x + t \cdot d^T x : x \in \mathbf{X},\ t \geq 0\}$ with $d := \mu^T D$.

For the solution of (10) an $O(\alpha \cdot T(m,n))$ algorithm is described in Chapter 5 where α denotes the number of breakpoints in the optimal value function and $T(m,n)$ stands for the complexity to solve the minimum-cost flow problem. In fact, we need not solve the whole parametric problem. Instead of the complete optimality function $x^*(t)$ we are searching for two special parameter values t_u and t_l such that

(11) $t_l = \arg\min \{t : \mu^T D \cdot x^*(t) \leq \mu^T d,\ x^*(t) \in \mathbf{Z}^m\}$ and
(12) $t_u = \arg\min \{t : D \cdot x^*(t) \leq d,\ x^*(t) \in \mathbf{Z}^m\}$.

While $x^*(t_u)$ delivers the best feasible solution (if there is some) for the original problem among all solutions described by $x^*(t)$, $x^*(t_l)$ is taken to compute an lower bound of z(P). In the case that $\mu^T D \cdot x^*(t_l) < \mu^T d$ it is possible that there is an integer solution x': $\mu^T D \cdot x' \leq \mu^T d$ and $c^T x' < c^T x^*(t_l)$.

From the equivalence between parametric and multicriteria programming it is known that the set of points $(c^Tx^*(t), d^Tx^*(t))$ forms a piece-wise linear convex function in the objective space. The solution procedure ANNA iteratively determines the breakpoints of this optimal value function starting with the solution of two lexicographical problems as defined in Chapter 5.

At each iteration, a certain weighted linear combination of c^Tx and d^Tx is taken to compute the next breakpoint or to show that two extreme point are neighbored. This procedure can be refined for our purposes by cutting off all $x^*(t)$: $d^Tx^*(t) > \mu^Td$ and also all extreme points which are feasible with respect to the additional constraints but worse than the solution of the actual best $x^*(t_u)$.

Doing in this way, our overall solution procedure called SURROGATE tries to improve the upper and lower bounds at each iteration. Concerning the 'best' choice of the step size β^k at the k-th iteration we present the final results after having completed the numerical study. There are three possibilities for the termination of the algorithm:

(i) The optimality conditions of Lemma 8.3. are fulfilled.
(ii) The difference between upper and lower bound does not exceed the value $\epsilon \cdot lb$.
(iii) The number of iterations is greater than an a priori given bound α.

```
procedure SURROGATE
begin
  ub := ∞
  lb := -∞
  k  := 1
  μ^k := d
  repeat
    compute x*(t) as a solution of S(μ^k,t) due to (10)
    compute t_u and t_l according to (11),(12)
    ub := min { ub,c^T x*(t_u) }
    lb := max { lb,c^T x*(t_l) }
    δ  := Dx*(t_l) - 1
    μ^{k+1} := μ^k + β^k δ
    k  := k+1
  until Dx*(t_l) ≤ d or (ub - lb) ≤ ε·lb or k < α
end
```

We illustrate the solution approach by the numerical example introduced above. The additional constraints are:

$x(4,17) + x(11,24) \le 1$
$x(8,21) + x(15,28) \le 1$
$x(9,22) + x(16,29) \le 1.$

With the initial dual vector $\mu^1 = (1,1,1)$ an optimal solution function $x^*(t)$ is obtained. This results in

$D \cdot x^*(t_1) = (2,1,0)^T$

and $lb = c^T x^*(t_1) = -8$, $ub = c^T x^*(t_u) = -7$.

In the second iteration, $\mu^2 := \mu^1 + (1/2)(D \cdot x^*(t_1) - 1) = (1.5,1,0.5)$ is used as dual multiplier. This results in a new optimality function and a new lower bound of -7. Consequently, an optimal solution with a maximum number of seven jobs has been found. The jobs are performed by the following machines:

machine 1 (class 1): jobs 8,9
machine 2 (class 1): job 3
machine 3 (class 2): jobs 1,2
machine 4 (class 2): jobs 6,4.

LIST of ALGORITHMS

LIST of PROBLEMS

REFERENCES

AASHTIANI, H.A.; MAGNANTI, T.L. (1976): Implementing primal-dual network flow algorithms. Technical Report OR 055-76, Operations Research Center, M.I.T., Cambridge, MA.

AGGARWAL, V.; ANEJA, Y.P.; NAIR, K. (1982): Minimal spanning tree subject to a side constraint. *Computers and Operations Research* 9, 287-296.

AHLFELD, D.P.; DEMBO, R.S.; MULVEY, J.M.; ZENIOS, S.A. (1987): Nonlinear programming on generalized networks. *ACM Transactions Math. Software* 13, 350-367.

AHRENS, J.H.; FINKE, G. (1980): Primal transportation and transshipment algorithms. *ZOR* 24, 1-72.

AHUJA, R.K.; GOLDBERG, A.V.; ORLIN, J.B.; TARJAN, R.E. (1988): Finding minimum cost flows by double scaling. To appear.

AHUJA, R.K.; MAGNANTI, T.L.; ORLIN, J. (1988): *Network Flows*. Sloan School of Management, M.I.T., Cambridge.

AHUJA, R.K.; ORLIN, J.B. (1986): A fast and simple algorithm for the maximum flow problem. Working Paper 1905-87, Sloan School of Management, M.I.T., Cambridge.

AHUJA, R. K.; ORLIN, J. B.; TARJAN, R. E. (1988): Improved bounds for the maximum flow problem. Research Report, Sloan School of Management, M. I. T., Cambridge.

AKGÜL, M. (1985a): Shortest path and simplex method. Research Report, Department of Computer Science and Operations Research, North Carolina State University, Raleeigh, N.C.

AKGÜL, M. (1985a): A genuinely polynomial primal simplex algorithm for the assignment problem. Research Report, Department of Computer Science and Operations Research, North Carolina State University, Raleeigh, N.C.

ALI, A.I.; CHARNES, A.; SONG, T. (1986): Design and implementation of data structures for generalized networks. *Journal of Optimization and Information Sciences* 8, 81-104.

ALI, A.I.; PADMAN, R.; THIAGARAJAN, H. (1989): Dual simplex-based re-optimization procedures for network problems. *Operations Research* 37, 158-171.

ANEJA, Y.P.; NAIR, K. (1979): Bicriteria transportation problem. *Management Science* 25, 73-78.

ARKIN, E.M.; SILVERBERG, E.L. (1987): Scheduling jobs with fixed start and end times. *Discrete Applied Mathematics* 18, 1-8.

BARAHONA, F.; TARDOS, E. (1988): Note on Weintraub's minimum cost flow algorithm. WP 88509 - OR, RFW Universität Bonn.

BARR, R. S.; FARHANGIAN, K.; KENNINGTON, J. L. (1986): Networks with side constraints: An LU factorization update. *The Annals of the Society of Logistics Engineers* 1, 66-85.

BARTHOLDI, J.J. (1982): A good submatrix is hard to find. *Operations Research Letters* 1, 190-193.

BELLING-SEIB, K.; MEVERT, P.; MUELLER, C. (1988): Network flow problems with one additional constraint: A comparison of three solution methods. *Computers and Operations Research* 15, 381-394.

BELLMAN, R.E. (1958): On a routing problem. *Quarterly Applied Mathematics* 16, 87-90.

BELLMAN, R.E.; ZADEH, L.A. (1970): Decision making in a fuzzy environment. *Management Science* 17, B141-B164.

BERTSEKAS, D.P. (1986): Distributed relaxation methods for linear network flow problems. *Proceedings of 25th IEEE Conference on Decision and Control*, Athens, Greece.

BERTSEKAS, D.P.; TSENG, P. (1988): Relaxation methods for minimum cost ordinary and generalized network flow problems. *Operations Research* 36, 93-114.

BIXBY, R.E. (1977): Kuratowski's and Wagner's theorems for matroids. *Journal of Combinatorial Theory* 22, Series B, 31-53.

BIXBY, R.E.; CUNNINGHAM, W.H. (1980): Converting linear programs to network problems. *Mathematics of Operations Research* 5, 321-357.

BIXBY, R.E.; FOURER, R. (1988): Finding embedded network rows in linear programs I: Extraction heuristics. *Management Science* 34, 342-375.

BIXBY, R.E.; WAGNER, D.K. (1985): An almost linear-time algorithm for graph realization. Technical Report 85-2, Rice University, Houston, Department of Mathematical Sciences.

BLAND, R.G.; JENSEN, D.L. (1985): On the computational behaviour of a polynomial-time network flow algorithm. Technical Report 661, School of Operations Research and Industrial Engineering, Cornell University, Ithaca, N.Y.

BLASCHKE, W. (1954): *Analytische Geometrie*. Birkhäuser Verlag, Basel, Stuttgart.

BOOTH, K.S.; LUEKER, G.S. (1976): Testing for the consecutive ones property, interval graphs and graph planarity using PQ-trees. *Journal Computer and System Science* 13, 335-379.

BRADLEY, G.H.; BROWN, G.G.; GRAVES, G.W. (1977): Design and implementation of large scale primal transshipment algorithms. *Management Science* 24, 1-34.

BROWN, G. G.; McBRIDE, R. D. (1984): Solving generalized networks. *Management Science* 30, 1497-1523.

BROWN, G. G.; McBRIDE, R. D.; WOOD, R. K. (1985): Extracting embedded generalized networks from linear programming problems. *Mathematical Programming Study* 32, 11-32.

BRUCKER, P. (1985): Parametric programming and circulation problems with one additional constraint. In: H. Noltemeier (ed.), *Proceedings WG'85*. Trauner Verlag, 27-41.

BRUCKER, P.; WIEMKER, T. (1989): NETGEN and BERTSEKAS (Pascal Codes).

European Journal of Operational Research 41, 102-103.

BURKARD, R.E.; HAMACHER, H.W.; ROTE, G. (1987): Approximation of convex functions and applications in mathematical programming. Report 89-1987, TU Graz, Institut für Mathematik.

BURKARD, R.E.; KRARUP, J.; PRUZAN, P.M. (1982). Efficiency and optimality in minimum, minimax 0-1 programming problems. *Journal Operational Research Society* 33, 137-151.

BUSACKER, R.G.; GOWEN, P.J. (1961): A procedure for determining a family of minimal-cost network flow patterns. Technical Paper 15, O. R. O., John Hopkins University, Baltimore.

CHANAS, S. (1983): The use of parametric programming in fuzzy linear programming. *Fuzzy Sets and Systems* 11, 243-251.

CHANDRU, V.; COULLARD, C.R.; WAGNER, D.K. (1985): On the complexity of recognizing a class of generalized networks. *Operations Research Letters* 4, 75-78.

CHARSTENSEN, P. (1983): Complexity of some parametric integer and network programming problems. *Mathematical Programming* 26, 64-75.

CHARSTENSEN, P. (1984): Parametric cost shortest chain problem. Bell Communications Research, Holmdel.

CHEN, S.; SAIGAL, R. (1977): A primal algorithm for solving a capacitated network flow problem with additional linear constraints. *Networks* 7, 59-79.

CHERKASSKY, R.V. (1977): Algorithm for construction of maximal flow in networks with complexity of $O(V^2 E)$ operation, *Mathematical Methods of Solution of Economical Problems* 7, 112-125 (in Russian).

CHEUNG, T. (1980): Computational comparison of eight methods for the maximum network flow problem. *ACM Transactions on Mathematical Software* 6, 1-16.

CHRISTOPHIDES, N. (1975): *Graph theory: An algorithmic approach*. Academic Press.

COOK, S.A. (1971): The complexity of theorem proving procedures. *Proceedings Third Annual ACM Symposium on Theory of Computing*, New York, 151-158.

CUNNINGHAM, W. H. (1976): A network simplex method. *Mathematical Programming* 11, 105-116.

CUNNINGHAM, W.H. ; EDMONDS, J. (1980): A combinatorial decomposition theory. *Canadian Journal of Mathematics* 32, 734-765.

DANTZIG, G.B. (1963): *Linear programming and extensions*. Princeton: Princeton University Press.

DANTZIG, G.B.; FULKERSON, D.R. (1954): Minimizing the number of tankers to meet a fixed schedule. *Naval Research Logistics Quarterly* 1, 217-222.

DERIGS, U.; MEIER, W. (1989): Implementing Goldberg's max-flow algorithm - A computational investigation. *ZOR* 33, 383 - 404.

DIAL, R.; GLOVER, F.; KARNEY, D.; KLINGMAN, D. (1979): A computational analysis of alternative algorithms and labeling techniques for finding shortest path trees. *Networks* 9, 215-248.

DIJKSTRA, E.W. (1959): A note on two problems in connection with graphs. *Numerische Mathematik* 1, 269-271.

DINIC, E.A. (1970): Algorithm for solution of a problem of maximum flow in networks with power estimation. *Soviet Math. Dokl.* 11, 1277-1280.

EDMONDS, J.; KARP, R.M. (1972): Theoretical improvements in algorithmic efficiency for network flow problems. *Journal of the ACM* 19, 248-264.

EISNER, M.J.; SEVERANCE, D.G. (1976): Mathematical techniques for efficient record segmentation in large shared databases. *Journal of the ACM* 19, 248-264.

ENGQUIST, M.; CHANG, M.D. (1985): New labeling procedures for the basis graph in generalized networks. *Operations Research Letters* 4, 151-155.

ELAM, J.; GLOVER, F.; KLINGMAN, D. (1979): A strongly convergent primal simplex algorithm for generalized networks. *Mathematics of Operations Research* 4, 39-59.

EVANS, J.R.; JARVIS, J.J.; DUKE, R.A. (1977): Graphic matroids and the multicommodity transportation problem. *Mathematical Programming* 13, 323-328.

FORD, L.R.; FULKERSON, D.R. (1956): Maximal flow through a network. *Canadian Journal of Mathematics* 8, 399-404.

FORD, L.R.; FULKERSON, D.R. (1962): *Flows in networks*. Princeton University Press, Princeton, New Jersey.

FREDMAN, M.L.; TARJAN, R.E. (1984): Fibonacci heaps and their uses in network optimization algorithms. *Proceedings 25th Annual IEEE Symposium on Foundations of Computer Science*, 338-346.

FRUHWIRTH, B.; BURKARD, R.E.; ROTE, G. (1989): Approximation of convex curves with application to the bicriterial minimum cost flow problem. *European Journal of Operational Research* 42, 326-338.

FUJISHIGE, S. (1980): An efficient PQ-graph algorithm for solving the graph realization problem. *Journal of Computer and System Sciences* 21, 63-86.

FUJISHIGE, S. (1986): An $O(m^3 \log m)$ capacity-rounding algorithm for the minimum-cost circulation problem: a dual framework of the Tardos algorithms. *Mathematical Programming* 35, 298-309.

FULKERSON, D.R. (1961): An out-of-kilter method for minimal cost flow problems. *SIAM Journal Applied Mathematics* 9, 18-27.

GABOW, H.N. (1985): Scaling algorithms for network problems. *Journal of Computer and System Science* 31, 148-168.

GABOW, H.N.; TARJAN, R.E. (1987): Faster scaling algorithms for network problems. To appear in SIAM Journal on Computing.

GALIL, Z. (1980) An $O(V^{5/3} \cdot E^{2/3})$ algorithm for the maximal flow problem. *Acta Informatica* 14, 221-242.

GALIL, Z.; NAAMAD, A. (1980): An $O(E \cdot V \cdot \log^2 V)$ algorithm for the maximal flow problem. *Journal of Computer and System Science* 21, 203-217.

GALIL, Z.; TARDOS, E. (1986): An $O(n^2(m+n \cdot \log n)\log n)$ min-cost flow algorithm. *Proceedings 27th IEEE Symposium on Foundations of Computer Science*, 1-9.

GALLO, G.; GRIGORIADIS, M.; TARJAN, R.E. (1989): A fast and simple parametric network flow algorithm and applications. *SIAM Journal on Computing* 18, 30-55.

GALLO, G.; PALLOTTINO, S. (1988): Shortest path algorithms. *Annals of Operations Research* 13, 3-79.

GALLO, G.; SODINI, C. (1979): Adjacent flows and application to min concave cost flow problems. *Networks* 9, 95-121.

GAREY, M.R.; JOHNSON, D.S. (1979): *Computers and intractability - A guide to the theory of NP-completeness*. Freeman, San Francisco.

GEOFFRION, A.M. (1968): Proper efficiency and the theory of vector maximization. *Journal of Mathematical Analysis and Applications* 22, 618-630.

GLOVER, F.; KLINGMAN, D. (1973): On the equivalence of some generalized network problems to pure network problems. *Mathematical Programming* 4, 269-278.

GLOVER, F.; KLINGMAN, D.; MOTE, J.; WITHMAN, D. (1979): Comprehensive computer evaluation and enhancement of maximum flow algorithms. Research Report CCS 356, Center for Cybernetic Studies, University of Texas, Austin, 1979.

GLOVER, F.; HULTZ, J.; KLINGMAN, D.; STUTZ, J. (1978): Generalized networks: a fundamental computer-based planning tool. *Management Science* 24, 1209-1220.

GLOVER, F.; KLINGMAN, D. (1981): The simplex SON algorithm for LP/embedded network problems. *Mathematical Programming Study* 15, 148-176.

programming problem to discrete generalized and pure networks. *Operations Research* 28, 829-836.

GOLDBERG, A.V. (1985): A new max-flow algorithm. Technical Report MIT/LCS/TM-291, Laboratory for Computer Science, M.I.T., Cambridge, MA.

GOLDBERG, A.V.; PLOTKIN, S.A.; TARDOS, E. (1988): Combinatorial algorithms for the generalized circulation problem. Research Report. Laboratory for Computer Science, M.I.T., Cambridge, MA.

GOLDBERG, A.V.; TARJAN, R.E. (1986): A new approach to the maximum flow problem. *Proceedings of the Eighteenth Annual ACM Symposium on the Theory of Computing*, 136-146.

GOLDBERG, A.V.; TARJAN, R.E. (1987): Solving minimum cost flow problem by successive approximation. *Proceedings of the Nineteenth Annual ACM Symposium on the Theory of Computing*, 7-18.

GOLDFARB, D.; GRIGORIADIS, M.D. (1988): A computational comparison of the Dinic and network simplex methods for maximum flow. *Annals of Operations Research* 13, 83-123.

GOMORY, R.E.; HU, T.C.(1961): Multi-terminal network flows. *Journal of SIAM* 9, 551-570.

GRIGORIADIS, M.D. (1986): An efficient implementation of the network simplex method. *Mathematical Programming Study* 26, 83-111.

GRIGORIADIS, M.D.; HSU, T. (1979): The Rutgers minimum cost network flow subroutines. *SIGMAP Bulletin of the ACM* 26, 17-18.

GROPPEN, V.O. (1987): Optimizing the realization of dynamic programming for a class of combinatorial problems. *Wissenschaftliche Zeitschrift der TH Leipzig* 11, 25-30.

GUSFIELD, D.M. (1980): Sensitivity analysis for combinatorial optimization. Memorandum UCB/ERL M80/22, University of California, Berkeley.

GUSFIELD, D.M. (1987): Very simple algorithms and programs for all pairs network flow analysis. Research Report No. CSE-87-1. Dept. of Computer Science and Engineering, University of California, Davis, CA.

GUPTA, U.L.; LEE, D.T.; LEUNG, J. (1979): An optimal solution to the channel assignment problem. *IEEE Transactions Comp.* C-28, 807-810.

HAMACHER, H.W. (1979): Numerical investigations on the maximal flow algorithm of Karzanov. *Computing* 22, 17-29.

HAMACHER, H.W.; FOULDS, L.R. (1989): Algorithms for flows with parametric capacities. *ZOR* 33, 21-37.

HANSEN, P. (1980): Bicriterion path problems. In: Multiple criteria decision making - Theory and applications (G. Fandel, T. Gal, eds.) *Lecture Notes in Economics and Mathematical System* Vol. 177, 109-127.

HOFFMAN, A.J.; KRUSKAL, J.B. (1956): Integral boundary points of convex polyhedra. In: H.W. Kuhn and A.W. Tucker (eds.): *Linear inequalities and related systems* 223-246. Princeton: Princeton University Press.

HUNG, M.S. (1983): A polynomial simplex method for the assignment problem. *Operations Research* 31, 595-600.

IGNIZIO, J.P. (1976): *Goal programming and extensions*. D.C. Heath, Lexington, MA.

IMAI, H. (1983): On the practical efficiency of various maximum flow algorithms. *Journal of the Operations Research Society of Japan* 26, 61-82.

IRI, M. (1968): On the synthesis of loop and cutset matrices and the related problems. *RAAG Memoirs*, Vol. 4, A-XIII, 4-38.

ISERMANN, H. (1982): Linear lexicographic optimization. *OR Spectrum* 4, 223-228.

JARVIS, J. J.; TUFEKCI, S. (1982): Decomposition algorithms for locating minimal cuts in a network. *Mathematical Programming* 22, 316-331.

KARWAN,, M.H.; RARDIN, R.L. (1979): Some relationship between Lagrangean and surrogate duality in integer linear programming. *Mathematical Programming* 17, 320-334.

KARZANOV, A. V. (1974): Determining the maximal flow in a network by the method of preflows, *Soviet Math. Dokl.* 15, 434-437.

KENNINGTON, J.L.; HELGASON, R.V. (1980): *Algorithms for network programming*. John Wiley and Sons, New York.

KENNINGTON, J.L.; MUTHUKRISHNAN, R. (1988); Solving generalized network problems on a shared memory multiprocessor. Technical Report 38-OR-21. Southern Methodist University, Department of Operations Research and Engineering Management.

KENNINGTON, J.L.; WHISMAN, A. (1988): NETSIDE user's guide. Technical Report 86-OR-01. Southern Methodist University, Department of Operations Research.

KLEIN, M. (1967): A primal method for minimal cost flows. *Management Science* 14, 205-220.

KLINGMAN, D.; NAPIER, A.; STUTZ, J. (1974): NETGEN: A program for generating large scale capacitated assignment, transportation, and minimum cost flow network problems. *Management Science* 20, 814-821.

KLINGMAN, D.; RUSSELL, R. (1975): On solving constraint transportation problems. *Operations Research* 23, 91-107.

KLINZ, B. (1988): Die Algorithmen von Goldberg und Orlin/Ahuja für das maximale Flussproblem. EDV-Projekt, TU Graz, Institut für Mathematik.

KLINZ, B. (1989): Maximale Flussprobleme und minimale Schnittprobleme mit affin linearen parametrischen Kapazitäten. Diplomarbeit, TU Graz, Institut für Mathematik.

KOENE, J. (1984): Minimal cost flow in processing networks: A primal approach. CWI Tract, vol. 4, Centrum voor Wiskunde en Informatica, Amsterdam.

KOEPPEN, E. (1980): Numerische Untersuchungen zum maximalem Flussproblem. Diplomarbeit, Universität zu Köln, Mathematisches Institut.

KOK, M.(1986): Conflict analysis via multiple objective programming, with experience in energy planning. Thesis, Delft University of Technology.

KOLEN, A.W.; KROON, L.G. (1989): On the computational complexity of (maximum-) class scheduling. Report nr. 46, Erasmus University Rotterdam, School of Management.

KOLEN, A.W.; LENSTRA, J.K.; PAPADIMITRIOU, C.H. (1988). Interval scheduling problems. Unpublished manuscript.

KROON, L.; RUHE, G. (1989): Solution of a class of interval scheduling problems using network flows. To appear in: *Proceedings of the 14th IFIP-Conference "System Modelling and Optimization"*, Leipzig 1989, Springer-Verlag.

LOOTSMA, F.A. (1988): Optimization with multiple objectives. Report 88-75, Faculty of Technical Mathematics and Informatics, Delft University of Technology.

MARTEL, C. (1987): A comparison of phase and non-phase network flow algorithms. Technical Report CSRL-87-2, Department of Electrical Engineering and Computer Science, University of California, Davis, CA.

McBRIDE, R.D. (1985): Solving embedded generalized network problems. *European Journal of Operational Research* 21, 82-92.

MALHOTRA, V.M.; KUMAR, M.P.; MAHESWARI, S.N. (1978): An $O(V^3)$ algorithm for finding maximum flows in networks. *Information Processing Letters* 7, 277-278.

MINTY,G.J. (1960): Monotone networks. *Proceedings Royal Society London*, 257 Series A, 194-212.

MULVEY, J. (1978): Pivot strategies for primal simplex network codes. *Journal of the Association for Computing Machinery* 25, 266-270.

MULVEY, J.; ZENIOS, S. (1985): Solving large scale generalized networks. *Journal of Information and Optimization Sciences* 6, 95-112.

MURTAGH, B.; SAUNDERS,M. (1978): Large-scale linearly constraint optimization. *Mathematical Programming* 14, 41-72.

MURTY, K. (1980): Computational complexity of parametric programming. *Mathematical Programming* 19, 213-219.

NOZICKA, F.; GUDDAT, J.; HOLLATZ, H.; BANK, B. (1974): *Theorie und Verfahren der parametrischen linearen Optimierung*. Akademie-Verlag Berlin (in German).

ONAGA (1967): Optimal flows in general communication networks. *Journal Franklin Institute* 283, 308-327.

ORLIN, J.B. (1984): Genuinely polynomial simplex and non-simplex algorithms for the minimum cost flow problem. Working Paper No. 1615-84, Sloan, M.I.T., Cambridge.

ORLIN, J.B. (1985): On the simplex algorithm for networks and generalized networks. *Mathematical Programming Study* 24, 166-178.

ORLIN, J.B. (1988): A faster strongly polynomial minimum cost flow algorithm. *Proceedings of the 20th ACM Symposium on the Theory of Computing*, 377-387.

ORLIN, J.B.; AHUJA,R.K. (1988): New scaling algorithms for the assignment and minimum cycle mean problem. Working Paper OR 178-88, Operations Research Center, M.I.T., Cambridge.

PICARD, C.; QUEYRANNE, M.(1980): On the structure of all minimum cuts in a network and applications. *Mathematical Programming Study* 13, 8 - 16.

REBMAN, K.R. (1974): Total unimodularity and the transportation problem: A generalization. *Linear Algebra and its Applications* 8, 11-24.

REINGOLD, E.M.; NIEVERGELT, J.; DEO, N. (1977): *Combinatorial algorithms*: Theory and Practice. Prentice-Hall, Englewood Cliffs, New Jersey.

RÖCK, H. (1980): Scaling techniques for minimal cost network flows. *Discrete Structures and Algorithms*. Ed. V. Page, Carl Hanser, München, 181-191.

ROTE, G. (1988): The convergence rate of the sandwich algorithm for approximating convex functions. Report 118-1988, Institut für Mathematik, TU Graz.

RUHE, G. (1985a): Detecting network structure in linear programming. 4. Jahrestagung der Forschungsrichtung "Optimierung auf Netzwerken und Netzplantechnik", Stadt Wehlen, Mai 1985. *Wissenschaftliche Berichte der TH Leipzig*, Heft 17, 34-42.

RUHE, G. (1985b): Characterization of all optimal solutions and parametric maximal flows in networks. *Optimization* 16, 51-61.

RUHE, G. (1987): Multicriterion approach to the optimal computer realization of linear algorithms. *Wissenschaftliche Zeitschrift der TH Leipzig* 11, 31-36.

RUHE, G. (1988a): Flüsse in Netzwerken - Komplexität und Algorithmen. Dissertation B, Sektion Mathematik und Informatik, TH Leipzig. (in German).

RUHE, G. (1988b): Complexity results for multicriterial and parametric network flows using a pathological graph of ZADEH. *ZOR* 32, 9-27.

RUHE, G. (1988c): Parametric maximal flows in generalized networks - Complexity and algorithms. *Optimization*, 19, 235-251.

RUHE, G.; FRUHWIRTH, B. (1989): Epsilon-Optimality for bicriteria programs and its application to minimum cost flows. Report 1989-140, Institut für Mathematik, TU Graz (appears in *Computing*).

SCHRIJVER, A. (1986): *Theory of linear and integer programming*. John Wiley & Sons.

SERAFINI, P. (1987): Some considerations about computational complexity for multiobjective combinatorial problems. In: Jahn,J.; Krabs, W. (Eds.), Recent advances and historical development of vector optimization. *Lecture Notes in Economics and Mathematical Systems* Vol 294, 222-232.

SEYMOUR, P.D. (1980): Decomposition of regular matroids. *Journal of Combinatorial Theory* (B) 28, 305-359.

SHILOACH, Y. (1978): An (n I $\log^2$ I) maximum-flow algorithm. Technical Report STAN-CS-78-802, Computer Science Department, Stanford University.

optimierung (in German). In: *Berichte des X. IKM*, Weimar 1984, Heft 5, 90-93.

SLEATOR, D.D.; TARJAN, R.E. (1983): A data structure for dynamic trees. *Journal Computer and System Science* 24, 362-391.

TARDOS, E. (1985): A strongly polynomial minimum cost circulation algorithm. *Combinatorica* 5, 247-255.

TARJAN, R.E. (1983): *Data structures and network algorithms*. Society for Industrial and Applied Mathematics, Philadelphia, PA.

TARJAN, R.E. (1984): A simple version of Karzanov's blocking flow algorithm. *Operations Research Letters* 2, 265-268.

TOMIZAWA, N. (1976): An $O(m^3)$ algorithm for solving the realization problem of graphs- a combinatorial characterization of graphic (0,1)-matrices (in Japanese). Papers of the Technical Group on Circuit and System Theory of the Institute of Electronics and Communication Engineers of Japan CST, 75-106.

TRUEMPER, K. (1976): An efficient scaling procedure for gain networks. *Networks* 5, 151-160.

TRUEMPER, K. (1977): On max flow with gains and pure min-cost flows. *SIAM Journal of Applied Mathematics* 32, 450-456.

TRUEMPER, K. (1985): A flexible decomposition approach for finding submatrices with certain inherited properties. Working Paper, University of Texas, Dallas.

TURING, A.M. (1936): On computable numbers, with an application to the Entscheidungsproblem. *Proceedings London Math. Society* 42, 230-265.

YOUNG, S.M. (1977): Implementation of PQ-tree algorithms. Masters Th., Department of Computer Science, University of Washington, Seattle Washington.

YU, P.L.; ZELENY, M. (1976): Linear multiparametric programming by multicriteria simplex methods. *Management Science* 23, 159-170.

ZADEH, N. (1973): A bad network problem for the simplex method and other minimum cost flow algorithms. *Mathematical Programming* 5, 255-266.

ZIONTS, S.(1985): Multiple criteria mathematical programming: an overview and several approaches. In: *Mathematics of multi-objective optimization*. CISM Courses and Lectures, No. 289, International Centre for Mechanical Science, Springer Verlag, Wien, 85 - 128.

ZIONTS, S.; WALLENIUS, J.(1976): An interactive programming method for solving the multiple criteria problem. *Management Science* 22, 652 - 663.

INDEX